The American-built M4 Medium Tank, christened Sherman by the British, served in every theatre of war from North Africa, Sicily and Italy, to north-western Europe and finally in the Far East. From late 1942 it was arguably most important armoured vehicle in Britain's arsenal, with the possible exception of the Churchill, and Field Marshal Montgomery, the overall commander of the British and Commonwealth units fighting in the Mediterranean and later in Normandy, considered the Sherman to be a universal tank, capable of both close infantry support and mobile warfare.

The basic design, a radical reworking of the unsatisfactory M3 project, underwent innumerable minor modifications and the final production figures exceeded 50,000 with the various models being built by eleven major manufacturers and a host of sub-contractors. Although this figure is impressive in itself it does not include the highly successful self-propelled artillery pieces or tank-destroyers which utilised the Sherman chassis. In addition, a British modification, generally referred to today as the Firefly, combined the Sherman tank's mechanical reliability with the killing power of the Ordnance Quick-Firing (QF) 17pdr gun to produce one of the most successful tank designs of the war (1).

The successful expulsion of the Axis powers from North Africa in May 1943 paved the way from an invasion of Sicily and ultimately the Italian mainland and in the long, bitter struggle that ensued the Sherman tank would play a major role. This theatre would also see the first employment of major armoured units from the nations of the British Commonwealth and it was in fact these units that modified the basic tank to fulfil the roles of specialist assault engineer vehicle and personnel carrier. Although the establishments of the Canadian, New Zealand and South African formations generally followed the British practice there were differences, sometimes significant, and I have endeavoured to highlight these in the text and the organisational charts that accompany the unit histories. In this context I would refer the reader to *Tank Craft 2: Sherman Tanks of the British Army and Royal Marines, Normandy Campaign 1944* which discusses the system of unit markings and other aspects of the British army of the late war period, particularly the structure of the armoured brigades and regiments, in some detail.

The Technical Details and Modifications section, which begins on page 49, cannot hope to be a definitive history of the development of the Sherman tank given the available space but hopefully it will serve as a useful spotters guide and also explain some of the terminology (2). The profiles of the Camouflage & Markings section and the comments on vehicle camouflage which begin on page 60 are largely based on my own findings and are sometimes at odds with acknowledged experts and where this is the case I have tried to make it clear. But before considering what are purely technical aspects some acknowledgement of the formations involved should be at least attempted.

## Notes

1. The reader should be aware that although I have used the term Firefly throughout this book to describe the 17pdr-armed Sherman tank it is unlikely that this term was in use during the war and is almost certainly a post-war sobriquet, since accepted into general usage.
2. Further explanation of the use of the Sherman by US forces can be found in *Tank Craft 11: Sherman Tanks US Army, North-Western Europe, 1944–1945.*

*A Sherman V of A Squadron, 14th Armoured Regiment (The Calgary Regiment), 1st Canadian Armoured Brigade photographed during the first week of September 1943 when the brigade landed at Reggio Calabria on the Italian mainland.*

Although the nations of the British Commonwealth had pledged their manpower resources to the war effort from the very first day of the conflict it was not until the latter half of 1942 that serious attempts were made to raise armoured formations, other than light reconnaissance units, to supplement the infantry and artillery formations which had by that stage seen considerable action. The prospect of large numbers of mechanically reliable American-built medium and light tanks becoming available due to the Lend-Lease programme was almost certainly a crucial factor in the decision to raise, equip and train Canadian, New Zealand and South African tank units in preparation for the coming battles on the European mainland. Below are brief histories of the formations involved and the armoured vehicles they were assigned. I should mention that only major battles and offensives are described here as it would be quite impossible to detail the thousands of skirmishes and firefights which these units were involved. I have used the term Firefly, almost certainly a post-war invention, purely as a matter of convenience to describe the 17pdr conversions based on the Sherman I and Sherman V tanks. The characteristics of these tanks and the other production models are explained in the Technical Details & Modifications section which begins on page 51. The organisation of British armoured regiments, and the terminology used during the Second World War period, is also covered in *Tankcraft 2: Sherman Tanks ,British Army and Royal Marines, Normandy Campaign 1944.*

**1st Canadian Armoured Brigade.** Originally raised in February 1941, this formation was one of only two Canadian independent armoured brigades to see action during the Second World War and was formed with a brigade headquarters, which was built around elements of The Prince Edward Island Light Horse, and three tank battalions created from The Ontario Regiment, The Three Rivers Regiment and The Calgary Regiment. All three were former militia units that had been reorganised as tank battalions in 1936 but at that time the armoured corps contained just a few Carden-Loyd machine gun carriers which were used for training.

The regiments were renamed in 1941 as 11th, 12th and 14th Army Tank Battalions respectively although they also retained their former titles, for example 11th Canadian Tank Battalion (The Ontario Regiment (Tank)) (1). Beginning in November 1941 the armoured regiments were equipped with British Churchill tanks and although these proved to be mechanically unreliable The Calgary Regiment employed theirs in the Dieppe raid of 19 August 1942 where most of the vehicles were abandoned on the beach.

In March 1943 the brigade began replacing the Churchills with Canadian-built Ram tanks in anticipation of a complete re-equip with the US-built M4 Sherman and by the following June the regiments had received their full allocation of the new tanks.

Unlike other independent armoured brigades this formation did not receive any light tanks for the reconnaissance troops, employing Universal Carriers instead, and did not contain and a motor infantry battalion. In mid-June 1943 the entire brigade began loading their tanks for the month-long journey to the Mediterranean where they were to take part in Operation Husky, the invasion of Sicily. Only The Three Rivers Regiment was involved in the initial landings, supporting 1st Canadian

........................text continued on page 4

### Notes

1. All were renamed as Tank Regiments by General Order 351/42 of 15 August 1942. The full regimental titles at this time were (Reserve) Army Tank Regiment, followed by the regiment's name, for example 11th (Reserve) Army Tank Regiment (The Three Rivers Regiment (Tank)). The Reserve designation was not dropped until 1 April 1946 and was included in the August 1943 name change mentioned below. These complicated titles were rarely used in contemporary documents and I have mentioned them here only for the sake of accuracy.

*A Sherman V of A Squadron, The Ontario Regiment photographed in the town of Trecastagni, north of Catania, during the fighting in Sicily in August 1943. Note the metal foot holds welded to the transmission cover. This tank is also shown on page 18 of the Camouflage & Markings section of this book.*

## 1ST CANADIAN TANK BRIGADE, JULY 1943

*Although detailed war establishments had been introduced and amended for Canadian armoured units as early as August 1940 this brigade was created using the special British tables of organisation for armoured formations operating in the Middle East. A new Canadian war establishment was instituted in January 1944 which raised the personnel strength and authorised eleven light tanks for the regimental reconnaissance troop and added an anti-aircraft troop with six purpose-built anti-aircraft tanks. But no attempt was made to reorganise the brigade until at least the following April and the anti-aircraft troops were dropped from all establishments later in the year. The tank strengths given below are of course the ideal and the actual numbers present when the brigade embarked for Sicily were four with the brigade headquarters, sixty-four with The Ontario Regiment, sixty-six with The Three Rivers Regiment and sixty-six with The Calgary Regiment. A total of forty-one tanks were held by the Tank Delivery Regiment. The brigade was renamed 1st Canadian Armoured Brigade in August 1943.*

**1st Canadian Tank Brigade Headquarters**

25th Tank De very Regiment (The Elgin Regiment)

Brigade Signals

No. 2 Light Field Ambulance, Royal Canadian Army Medical Corps

83 Company, Royal Canadian Army Service Corps

*The headquarters also controlled 1st Canadian Army Tank Troops Workshop, 1st Canadian Heavy Recovery Company and the brigade ordnance company. 1st Canadian Assault Troop, a specialist engineer unit, did not join the brigade until June 1944.*

11th Tank Regiment (The Ontario Regiment)

Headquarters Squadron

*In addition to the four tanks, each regimental headquarters also contained a reconnaissance troop equipped with six scout cars and ten Universal Carriers, an intercommunication troop with eight scout cars and an administrative troop. No.59 Light Aid Detachment was on temporary assignment to the regiment.*

A Squadron

Administrative Troop

Squadron Headquarters

1 Troop

2 Troop *As for 1 Troop*

3 Troop *As for 1 Troop*

4 Troop *As for 1 Troop*

5 Troop *As for 1 Troop*

B Squadron *As for A Squadron*

C Squadron *As for A Squadron*

*Troops were numbered consecutively throughout the regiment*

12th Tank Regiment (The Three Rivers Regiment)

Headquarters Squadron

*In addition to the four tanks, each regimental headquarters also contained a reconnaissance troop equipped with six scout cars and ten Universal Carriers, an intercommunication troop with eight scout cars and an administrative troop. No.60 Light Aid Detachment was on temporary assignment to the regiment.*

A Squadron

Administrative Troop

Squadron Headquarters

1 Troop

2 Troop *As for 1 Troop*

3 Troop *As for 1 Troop*

4 Troop *As for 1 Troop*

5 Troop *As for 1 Troop*

B Squadron *As for A Squadron*

C Squadron *As for A Squadron*

*Troops were numbered consecutively throughout the regiment*

14th Tank Regiment (The Calgary Regiment)

Headquarters Squadron

*In addition to the four tanks, each regimental headquarters also contained a reconnaissance troop equipped with six scout cars and ten Universal Carriers, an intercommunication troop with eight scout cars and an administrative troop. No.61 Light Aid Detachment was on temporary assignment to the regiment.*

A Squadron

Administrative Troop

Squadron Headquarters

1 Troop

2 Troop *As for 1 Troop*

3 Troop *As for 1 Troop*

4 Troop *As for 1 Troop*

5 Troop *As for 1 Troop*

B Squadron *As for A Squadron*

C Squadron *As for A Squadron*

*Troops were numbered consecutively throughout the regiment*

.........................text continued from page 4

Infantry Brigade and 51st Highland Division in the assault on the beaches around Pachino on 10 July 1943, catching the Germans largely by surprise. Three days later the remainder of the brigade began arriving and for the rest of the campaign the role of the brigade headquarters, The Ontario Regiment and The Calgary Regiment would be that of fire support under the direct command of British XIII Corps.

The Three Rivers Regiment remained with the Canadian infantry, taking part in the capture of Modica and Vizzini, where they came up against German armoured units for the first time, and the bitter struggle for Leonforte. At the conclusion of twenty-eight days fighting the regiment had lost seven officers and 76 other ranks killed and wounded and almost a complete company of Sherman tanks, either destroyed or relinquished to rear area workshops. Most of the regiment's tanks had suffered some kind of battle damage.

In late August 1943, while preparing for Operation Baytown, the invasion of the Italian mainland, the brigade was redesignated 1st Canadian Armoured Brigade, although the regiments were referred to in official documents as Tank Regiments for some time (1).

On 3 September the brigade took part in the landings near Reggio Calabria in southern Italy with a Sherman V tank named Arrant of A Squadron, 14th Armoured Regiment claiming to be the first Allied tank to return to the European continent. A photograph of this tank is shown below.

The brigade was engaged in the fighting for Potenza, Termoli, and famously Ortona which the Canadian official history referred to as 'Little Stalingrad'. During this time Brigadier Robert A. Wyman, who had led the brigade since January 1942, left to take command of 2nd Canadian Armoured Brigade and was replaced by Brigadier William C. Murphy, a former militia officer who had accepted a reduction in rank to serve overseas.

In May 1944, during the fourth and final battle for Cassino, 1st Canadian Armoured Brigade helped break the German defences of the Gustav Line, pushing on with the infantrymen of 8th Indian Division to secure a crossing of the Garigliano River, south of Rome. Throughout the latter part of 1944 the brigade was attached to 1st Canadian Division and then British 78th Division during the offensives to break the Hitler Line (2).

The brigade was later subordinated to British XIII Corps, taking part in the battles around Lake Trasimene and the advance to the Gothic Line, the Germans' last major defensive positions in northern Italy. At the end of the year the Canadian units were withdrawn from the front and in early 1945 left Italy to join First Canadian Army in the Netherlands.

## Notes

1. General Order 88/44 of 26 August 1943 instituted the name change from Tank to Armoured but the new designation was adopted in the field until at least November.
2. The Germans very quickly changed the name to the Senger Line to avoid any negative fallout should the defences not hold. The first Allied unit to breach the line was in fact the 4th Princess Louise Dragoon Guards, a Canadian regiment.

*The Sherman V of A Squadron, The Calgary Regiment named Arrant mentioned in the text. Another photograph of this tank is shown in the Technical Details and Modifications section on page 61 where the camouflage scheme is discussed.*

*Throughout its service in Sicily and Italy the headquarters and regiments of 1st Canadian Armoured Brigade were equipped with Sherman V 75mm-armed tanks and as late as September 1944 this was the only type on hand. At left is shown a Sherman V from the regimental headquarters squadron of The Three Rivers Regiment photographed as it comes ashore near Pachino in Sicily in July 1943. Although this image is of mediocre quality I have included it here as it shows the rarely seen squadron tactical sign with the letter R which is repeated on the turret side. Note also the waterproof cover for the gun mantlet and hull machine gun, the .50 calibre machine gun mounted on the commander's cupola and the aerial pennants.*

*Deliveries of 17pdr-armed tanks began in late 1944 and these were all assigned to The Calgary Regiment. By the end of 1944 the regiment had received eighteen Sherman IB models equipped with the 105mm howitzer and these were distributed among the three regiments. Just prior to the brigade's move to North-west Europe further deliveries of 17pdr armed tanks were received and assigned to the other regiments. The Sherman IC shown at right is from The Ontario Regiment and was photographed in March 1945 as it was transported by rail to Belgium. Seated for left to right are Troopers Gyles, Simmons, Cartner and Clark.*

**5th Canadian Armoured Brigade.** Formed by renaming the 2nd Canadian Armoured Brigade in January 1943, this unit was commanded by Brigadier George R. Bradbrooke and attached to 5th Canadian Armoured Division (1).

The brigade contained a headquarters, and three armoured battalions that had been raised from Lord Strathcona's Horse (Royal Canadians), 8th Princess Louise's (New Brunswick) Hussars and The British Columbia Dragoons. These units were numbered as the 2nd, 5th and 9th Armoured Regiments respectively. As with the independent 1st Canadian Armoured Brigade, the regiments were referred to by the new number and their former titles, such as 9th Armoured Regiment (The British Columbia Dragoons) (2).

The brigade's motor infantry battalion was built around The Westminster Regiment but this was later transferred to 12th Infantry Brigade. The division trained for two years in the UK before moving to Italy in November 1943 as part of I Canadian Corps, leaving all its tanks and heavy equipment in Britain. It was planned that the Canadians would take over the tanks and transport vehicles of British 7th Armoured Division but it soon became clear that most of these were either worn out, having served in North Africa, or useless. The division's commander, Major-General Guy Simonds, was given the choice of accepting a number of Sherman III tanks that were currently held in Egypt or waiting for a complete shipment of Sherman V vehicles which were in transit from the UK. Reasoning that parts for the diesel-engined Sherman III models would be difficult to procure, Simmonds opted to wait for the Sherman V shipment when he was promised that they would arrive in Italy in less than a month. But by the end of the year just a handful of tanks had arrived and these were all assigned to The Governor General's Horse Guards, the division's armoured reconnaissance regiment (3).

These were in fact Sherman II models and as 246 of this type were in transit from North America in late 1944 the regiment's allocation probably consisted of new tanks. Two of these are depicted in the Camouflage & Markings section on page 20 and the probability they they were not former 8th Army vehicles is suggested by the later production features visible on both. I should mention here that the often quoted assertion that the division was equipped with a number of Sherman III tanks is not correct.

The 15th Army Group tank strength returns, which are quite detailed, show that both Canadian armoured brigades that operated in Italy were equipped with Sherman V tanks, except the Sherman II vehicles mentioned above, and these were later augmented by small numbers of Sherman IB and Sherman IC Fireflies.

## Notes

1. The division itself had been created from redesignated elements of 1st Canadian Armoured Division.
2. Unlike 1st Armoured Brigade the term Tank was never employed, all regiments being referred to as Armoured from April 1941. The Princess Louise's Hussars and The British Columbia Dragoons regiments retained the designation Reserve in their titles until 1946, for example 5th (Reserve) Armoured Regiment, 8th Princess Louise's (New Brunswick) Hussars.
3. This unit was organised along the same lines as the division's armoured regiments. In fact it was originally designated 3rd Armoured Regiment in February 1941 with the title changing to 3rd Armoured Reconnaissance Regiment (The Governor General's Horse Guards) on 1 January 1943. In a number of histories it is incorrectly assumed to be part of 5th Armoured Brigade

Deliveries of 17pdr-armed tanks began on 4 October 1944 and the first four, all Sherman IC models, were trialled by Lord Strathcona's Horse in an action fought between San Marino and Forli later the same month. At that time the tanks were all concentrated in one squadron and assigned to the troop sergeants but as further replacements arrived one 75mm-armed tank in each troop was replaced by a 17pdr-armed Sherman tank. This was the only regiment of the brigade to operate the Firefly until the division arrived in northern Europe.

The eighteen Sherman IB models which were eventually assigned to the brigade were distributed evenly to the three regiments where they were allotted to each squadron headquarters troop.

It took several months for the division to be fully equipped and only 11th Canadian Infantry Brigade was committed to battle before the spring of 1944. But the armoured units were ready to take part in the assaults on the German positions around Monte Cassino and they were instrumental in breaking the Hitler Line in May 1944. In the following August the tanks of 5th Canadian Armoured Brigade, now led by Brigadier James D. B. Smith, were engaged in Operation Olive, including the gruelling battle for Coriano where the Canadian tanks supported British 1st Armoured Division, and the advance to the Gothic Line, the last major German defensive works in northern Italy.

At the end of the year all Canadian ground units were withdrawn from Italy and in early 1945 joined First Canadian Army in the Netherlands. The organisation of this division and also the armoured brigade is examined on page 8.

........................*text continued on page 9*

*Major-General Bert Hoffmeister, who took command of 5th Canadian Armoured Division on March 1944, in the cupola of his Sherman V named Vancouver. Note the additional antenna in its armoured cover in front of the lap gunner's position. This tank is also shown in the Camouflage & Markings section on page 19 and a partial colour image is reproduced on the back cover.*

Said to have been photographed when 5th Canadian Armoured Division was advancing towards the Gothic Line defences in August 1944, this Sherman II may be one of the tanks assigned to the division's 3rd Armoured Reconnaissance Regiment (The Governor General's Horse Guards).

Photographed in March 1944, this tank of 5th Armoured Regiment (8th Princess Louise's (New Brunswick) Hussars) is fitted with the twin Bren light machine gun mount which was probably a field modification unique to this regiment. This image also provides a good view of the standard radio antenna mounts.

## 5TH CANADIAN ARMOURED DIVISION, SUMMER 1944

*Due to manpower restrictions and the anticipated need to promote cooperation between Canadian and British armoured units, a complete reorganisation was instituted for Canadian formations serving overseas on 11 January 1943. These new tables of organisation replaced those of September 1941 and, with minor exceptions, conformed to the British war establishment for an armoured regiment allowing for sixty-nine cruiser tanks, including one command tank, ten Universal Carriers, and nine scout cars. Each regiment contained three squadrons and each squadron was made up of five troops. But by early May 1943 this establishment had been replaced to reduce the number of cruiser tanks to sixty-one and introduce eleven light tanks and one armoured recovery tank for each squadron. The personnel strength was complicated by the allowance for five-man, six-man and seven-man crewed vehicles. With the advent of the Sherman tank as the standard cruiser design a new establishment was introduced in January 1944 and it is this organisation that was used by 5th Canadian Armoured Division in Italy and is depicted below. Unlike British divisions the reconnaissance unit was a fully-equipped armoured regiment and this allowed the division, with some very temporary reorganisation, to field the equivalent of two independent armoured brigade groups. During the summer battles of 1944 the division was equipped throughout with Sherman V and Sherman II models and M3/M5 Stuart light tanks but as late as September the regiments were all still under strength. It should also be mentioned that, at different times, some Canadian armoured regiments adopted an unofficial organisation where each squadron headquarters was equipped with three Sherman tanks in place of four while the squadrons contained four troops, instead of the authorised five. Each troop was reorganised with an additional tank, giving a total of four. This was purely a tactical innovation and did not change the overall number of vehicles. In October 1944 17pdr-armed Sherman Fireflies began to arrive and the first four of these, all Sherman IC versions, were trialed by Lord Strathcona's Horse.*

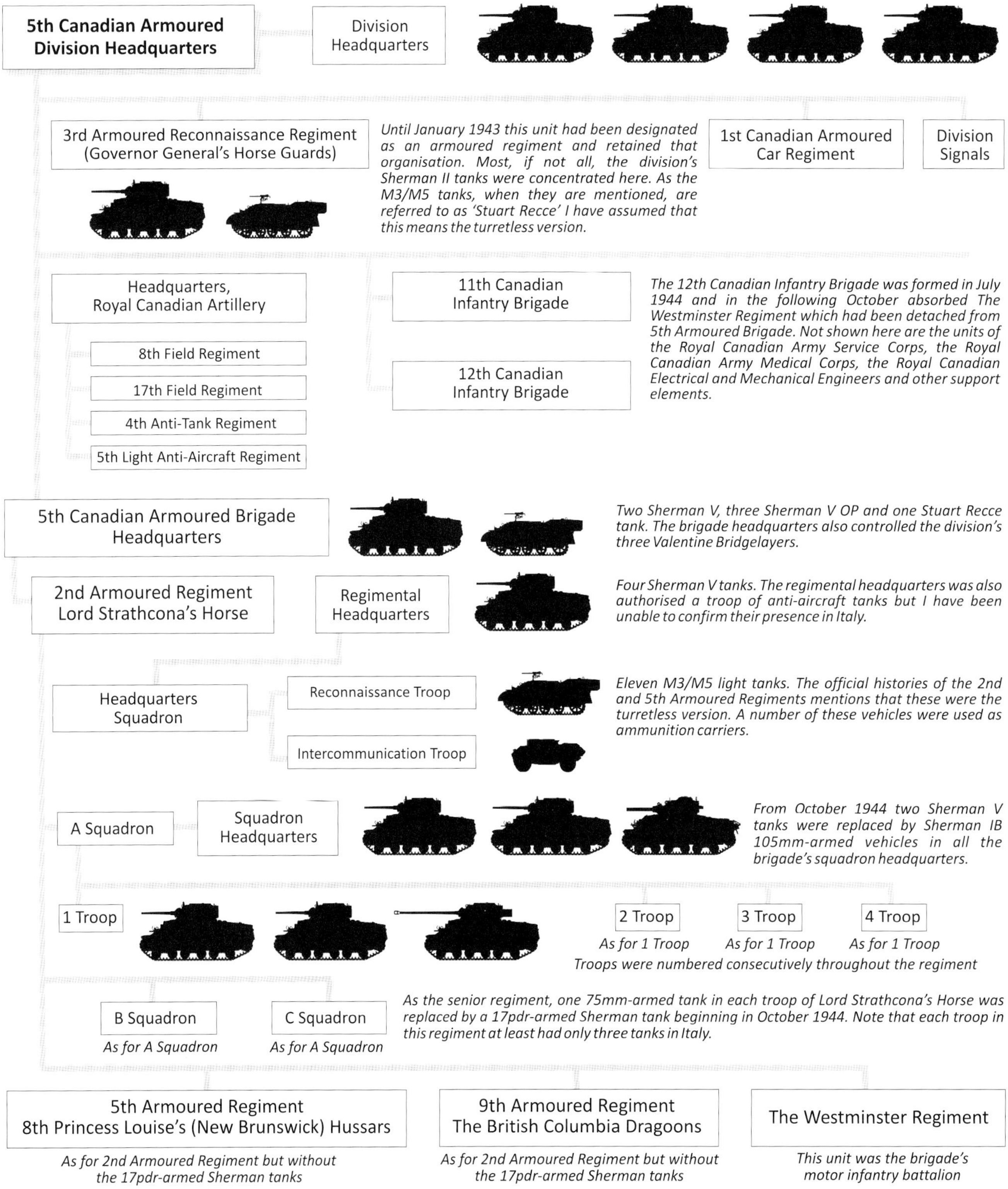

........................*text continued from page 6*

**4th New Zealand Armoured Brigade.** Formed in October 1942, the brigade contained a headquarters and three tank battalions referred to as the 18th, 19th and 20th Armoured Regiments. All three were formed from infantry battalions which had served in the Western Desert with replacements made up from an armoured brigade that was in the process of formation in New Zealand. While the regiments were training in Egypt, 22nd Motor Battalion joined the brigade (1).

Attached to 2nd New Zealand Division, the brigade was commanded by Brigadier Lindsay Merritt Inglis, an experienced soldier who had served in the First World War and throughout the North African campaign. The armoured regiments were each organised with fifty-two cruiser tanks, along the same lines as British 8th Army units of the time.

Each regimental reconnaissance troop was initially equipped with eleven Daimler scout cars but these were replaced in mid-1944 with Stuart V light tanks. Of limited value in a fight, the turrets were removed from the majority, usually as many as eight, and this allowed some to be used as ammunition carriers. This is explained in detail in the chart on page 11.

Throughout the Italian campaign the regiments of this brigade were equipped with diesel-engined Sherman III models until the arrival of a number of 17pdr-armed tanks in late 1944 and Sherman IB howitzers in early 1945 (2).

Twelve Fireflies were received in October 1944 and four were assigned to each regiment, meaning that all squadrons possessed at least one 17pdr-armed tank. The Firefly conversions were initially Sherman IC models but most had been replaced by Sherman VC versions by the spring of 1945. The 15th Army Group tank strength report of 7 April shows that twenty-four of the latter were on hand with the brigade at that time together with 122 Sherman III models. In addition, a total of 102 Sherman VC were either in stock or in transit from the UK and it is likely that the brigade received some of these, probably as many as twelve, bringing the total to thirty-six. This allowed 18th Armoured Regiment and 19th Armoured Regiment to equip each troop with one Firefly while 20th Armoured Regiment created a Firefly

## Notes

1. There is some confusion about when this infantry formation was reorganised as a motorised battalion. In any case, it left the brigade in October 1944.

2. The New Zealanders were famous for repairing and employing numbers of abandoned US Army and British vehicles. Sherman II tanks, both the so-called small and large hatch versions, and Sherman V Dozers were on hand in 1945 with 28 Assault Squadron, a combat engineer unit. A photograph of two of the squadron's fascine carriers based on Sherman II hulls is reproduced on page 16.

*A Sherman III of 20th Armoured Regiment, 4th New Zealand Armoured Brigade photographed at around the time of the Cassino battles in mid-1944. Note the cast driver's and lap gunner's hoods with their angled sides, the opened hatches with periscopes and the early gun mantlet.*

*A Sherman III of 19th Armoured Regiment, 4th New Zealand Armoured Brigade photographed in the devastated town of Cassino on 18 May 1944, the day the last German positions fell. Note the C Squadron tactical sign enclosing the troop number on the hull side and painted onto the ammunition box at the rear. Note also the Crusader-style blanket box and exhaust deflector.*

troop attached to the regimental headquarters and also placed one 17pdr-armed tank in each squadron headquarters. The same report shows that eighteen 105mm-armed Sherman IB were on hand and two of these were initially allocated to each squadron and distributed randomly as though they were standard 75mm-armed tanks. This arrangement proved problematic and was later altered to concentrate all the Sherman IB vehicles in the squadron headquarters troop. The exception to this rule was 19th Armoured Regiment which raised a special troop consisting of one Firefly, one 75mm-armed tank and two Sherman IB.

The brigade arrived in Italy in October 1943 and was involved in the attempts to break the defences of the Bernhardt Line along the Sangro River and later in the fighting for Cassino where the tanks fought their way through the streets of the town. In the autumn of 1944 the brigade took part in the attacks on Monte Castelle where the Shermans of 18th Armoured Regiment were used to tow the division's anti-tank guns and their crews to the front. Slowed by the many rivers of northern Italy and the deteriorating weather, most 8th Army tank units were employed as mobile artillery until winter effectively brought a halt to any offensive operations. In early April the Allies launched Operation Grapeshot which pushed through the Lombardy Plain past Bologna and Verona and east to Treviso, stopping only when the Germans surrendered on 2 May 1945.

Elements of 4th New Zealand Armoured Brigade advanced as far as Trieste on the Adriatic coast where they were welcomed by an enthusiastic crowd of Italian locals and surrendering German troops. But in their final action of the war the men of 19th Armoured Regiment endured a tense stand off with about twenty-five T-34 tanks manned by Yugoslav communist partisans, who had occupied part of the city, in what could perhaps be described as an early preview of the Cold War.

Plans were made to send a New Zealand armoured force to the Pacific but these were shelved with the Japanese surrender and on 2 December 1945 the 4th New Zealand Armoured Brigade was officially disbanded.

........................*text continued on page 12*

## 2ND NEW ZEALAND DIVISION, SUMMER 1944

*When the division arrived in Italy in October 1943 the tank component, 4th Armoured Brigade, was organised as shown below. In addition to the brigade's Sherman III tanks the Divisional Protective Troop, which was attached to the division headquarters, had at least three Shermans on hand. This unit was disbanded in June 1944 and its personnel and vehicles allocated to the regiments of the armoured brigade. In late 1944 enough Sherman Firefly 17pdr tanks were received to allow one to be allocated to each squadron and by March 1945 the brigade had a total of thirty-six on hand. This enabled 18th Armoured Regiment and 19th Armoured Regiment to assign one to each troop while 20th Armoured Regiment grouped theirs into a single troop attached to the regimental headquarters. Also in 1945, 105mm-armed Sherman IB tanks began arriving and two of these were assigned to each squadron except in 19th Armoured Regiment where a special troop was created with one 75mm-armed Sherman III, two Sherman IB tanks and a single Firefly.*

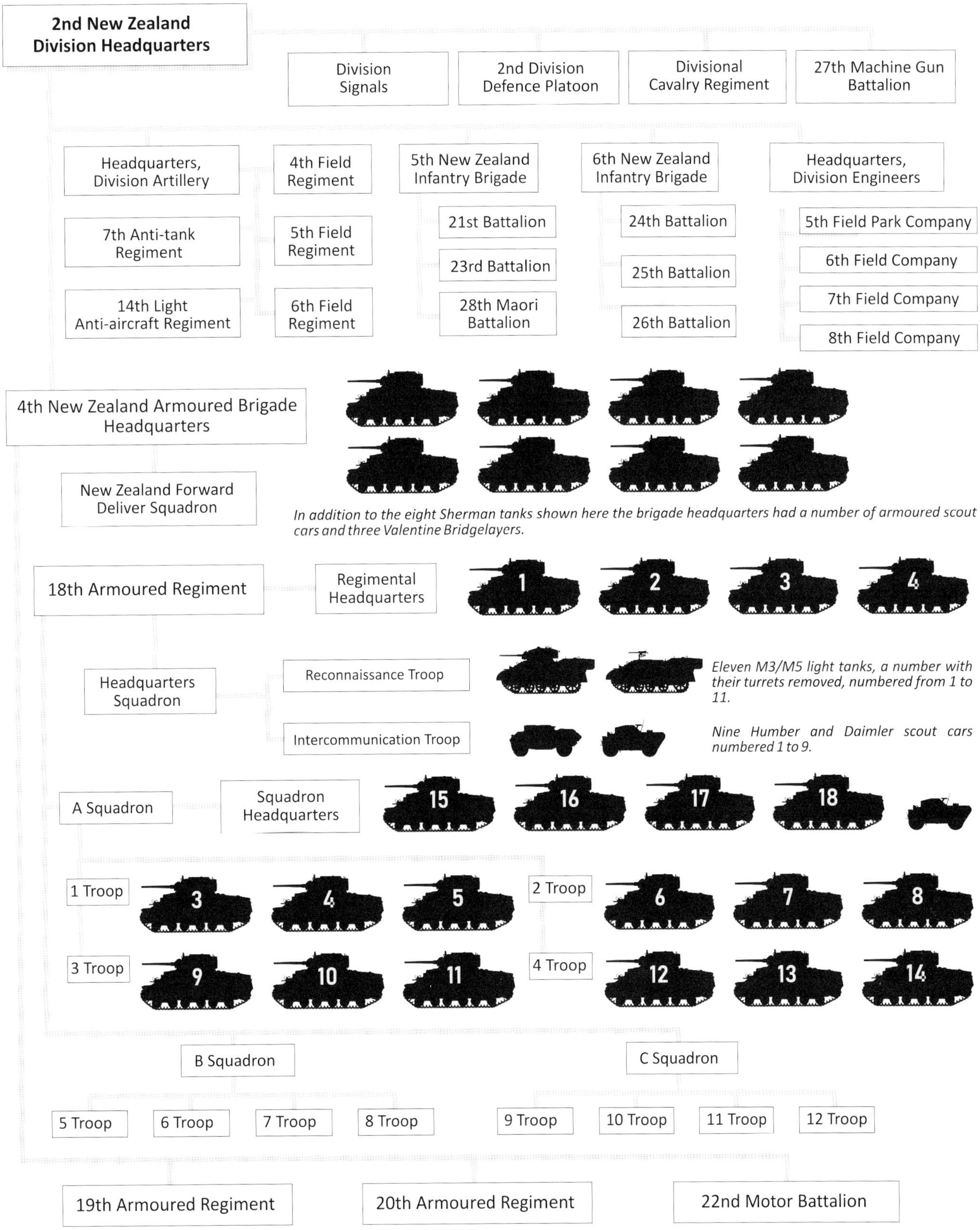


*The numbering system shown here was supposed to have been used throughout the brigade but its adoption seems to have been slow and in some instances it was not applied to the hull side as authorised but rather to the turret or some other surface. By late 1944 many tanks also displayed the troop number, usually inside the squadron tactical sign, but again this was not universal and some vehicles placed the tank's individual number, as shown above, there. Examples are shown in the Camouflage & Markings section.*

.........................text continued from page 10

**11th South African Armoured Brigade.** Formed in South Africa on 1 February 1943, the brigade was attached to 6th South African Armoured Division. Together with 12th South African Motor Brigade, the only other element of the division to be raised at that time, the brigade was shipped to Egypt in the following April.

The armoured brigade was made up of a headquarters with The Pretoria Regiment (Princess Alice's Own), Prince Alfred's Guards and the Special Service Battalion forming the armoured regiments. The division trained at Khataba, north-west of Cairo, and was steadily reinforced with South African and Rhodesian volunteers until 3 March 1944 when the division's commander, Major-General William Poole, received orders that his division would move to Palestine for further training.

But within days these orders had been countermanded and 6th South African Division was sent to Italy, the first units arriving in Taranto on 14 April 1944. At this time the division was almost at full strength but the organisation of its armoured brigade was slightly confused as is explained in the chart detailing the division's establishment on page 14 (1). In addition to the tanks of 11th South African Armoured Brigade the division's reconnaissance regiment, The Natal Mounted Rifles, was equipped as an armoured regiment and had forty-three Sherman and thirty Stuart V light tanks on hand when the division arrived in Italy. Sherman Artillery Observation Post (AOP) tanks were assigned to the division's artillery regiments and these are also mentioned on page 14 (2). All the 75mm-armed vehicles and AOP tanks were Sherman V models.

Commanded by Brigadier Jan P. A. Furstenburg, a former air force pilot, the brigade first saw action in the battle for the town of Celleno on 10 June 1944 and the tanks acquitted themselves well, opening the road to Florence. In the early weeks of July the division played an important role in the capture of Chianciano Terme and Montepulciano, west of Lake Trasimene, and the important railway junction of Castelinuovo Berardenga. By 3 August 1944 advance elements of the division had entered Florence.

On 26 August 1944, by now subordinated to US Army II Corps, the South African Division crossed the Arno River against light German opposition and pushed northward to reach Empoli, after crossing the river a second time. On 11 September the South Africans came up against the defences of the Gothic Line outside the city

## Notes

1. The number of Sherman tanks on hand would suggest that the Prince Alfred's Guards regiment was organised with five-troop squadrons, which would normally give a total of sixty-one tanks, while the squadrons of the two other regiments contained four troops, meaning that each regiment contained fifty-two tanks. Assuming this is correct, all three regiments were slightly over strength when they left Egypt.

2. Strictly speaking, Sherman AOP tanks were fitted with a dummy wooden gun to allow for the installation of additional radio equipment and a map board inside the turret. Photographs suggest that some Shermans retained their 75mm weapon and these vehicles were almost certainly field conversions.

*A Sherman V of 11th South African Armoured Brigade photographed near Monzuno, south of Bologna, on 28 December 1944. The turret and hull show remnants of the Light Mud and Black camouflage scheme that was typical of this brigade's tanks when they arrived in Italy earlier in the year but the applique armour plates seem to have been a later addition and one features a gun calibration marking. Although this vehicle is usually associated with The Pretoria Regiment, the C Squadron tactical sign on the hull is very light and it may have been rendered in the green which was sometimes used by the Special Service Battalion.*

*Photographed on 14 April 1945 in northern Italy, just weeks befor the end of the war, this Sherman IIA is one of the tanks assigned to 6th South African Armoured Division, identified by the division's formation badge which is just visible on the left side of the transmission housing. Note the lengths of wire strung across the hull and turret to hold foliage camouflage. The name on the hull side is unfortunately indecipherable.*

of Lucca on the Serchio river, about 50 kilometres west of Florence, but here they were held up for two weeks before taking part in an attack on the German positions around Monte Stanco north of Pistoia. By now the Allied armies were exhausted and, succumbing to the appalling weather, the advance ground to a halt. On 26 October 1944 the South African division was withdrawn from the front line and went into winter quarters near the town of Veggio.

The division returned to the fighting on 15 April 1945, taking part in an attack along the Florence to Bologna railway line, and as the Germans were now in headlong retreat, the mobile elements were used to full effect, capturing the Panaro River bridge east of Camposanto, between Modena and Ferrara, intact. The tanks pushed forward, reaching Bondeno by Monday, 23 April and established a bridgehead over the Po River on the following Wednesday near the village of Felonica, north-west of Ferrara. By this time German soldiers were surrendering in large numbers, in part to escape the attention of Italian partisans. To oppose an expected German counteroffensive the division was moved to the area around Milan where, on 2 May 1945, the South Africans heard that the war was over. The South African armoured brigade was numerically the most important user of the 76mm-armed Sherman IIA, a model which the British Army had accepted only grudgingly. Numbers of these tanks began arriving in Italy in September 1944 and at the end of the month the South Africans reported that fourteen were on hand. But following the enforced winter hiatus, this number had increased dramatically to 129 as reported in February 1945. The South African armoured brigade was in fact one of the most technically diverse units in Italy with numbers of Sherman V, Sherman IB, Sherman IC Fireflies and Sherman IIA tanks on hand, the latter being the most numerous during the final weeks of the conflict.

Unfortunately the last 15th Army Group strength report available to us today is that dated 7 April 1945 and although no Sherman VC Fireflies are recorded as being assigned to the brigade, photographic evidence shows that some were on hand with The Pretoria Regiment at least (1). On 13 April 1945, Major-General Poole wrote that a number of the division's 75mm-armed Sherman tanks had been 'exchanged' for nine Sherman IB, nine 17pdr-armed Shermans, six Sherman Bulldozers and three Sherman Flails, the latter presumably Sherman Crabs. This exchange of tanks may explain the presence of the Sherman VC models.

## Notes

1. A photograph of these tanks is reproduced on page 59 and one is also depicted in the Camouflage & Markings section on page 25. The suggestion that the vehicles have some connection to Operation Olive, the first attempt to break the Gothic Line in August 1944, is incorrect.

## 6TH SOUTH AFRICAN DIVISION, SPRING 1944

*A great deal of contradictory information exists regarding the organisation and equipment of this division and particularly 11th South African Armoured Brigade. One repeatedly quoted source states that significant numbers of Sherman VC 17pdr-armed tanks and Sherman Armoured Recovery Vehicles (ARV) were on hand with the brigade by August 1944 when no examples of these vehicles were in Italy at that time. Similarly, the suggestion that each regiment contained four fully-equipped tank squadrons during the summer of 1944 is unlikely to be correct as the division's official embarkation list compiled on 27 May 1944 shows that most would have been hard pressed to fill three squadrons. A comprehensive study of the brigade's baptism of fire, the Battle of Celleno which took place on 10 June 1944, consistently mentions that the regiments were made up of three squadrons (1). To further confuse the issue, the May 1944 totals would suggest that The Pretoria Regiment and The Special Service Battalion were organised on a four-troop basis, as most 8th Army regiments had been, while the squadrons of Prince Alfred's Guard each contained five troops. The latter establishment was commonly found in Italy among British units which had been part of 1st Army in Tunisia. The South African division underwent a number of internal organisational changes between May and December 1944 and that regiments did in fact contained a D Squadron is confirmed by photographic evidence. But I feel that this was a later development made possible by the arrival of Sherman VC and Sherman IB tanks in 1945. As mentioned earlier, the numbers given below are derived from the May 1944 shipping lists, rendered in a lighter shade, and also the September 1944 and February 1945 strength returns prepared by 15th Army Group headquarters. Where I have been forced to rely on secondary sources or to speculate I have tried to make this clear. Note that the division's infantry component and support units are not shown here.*

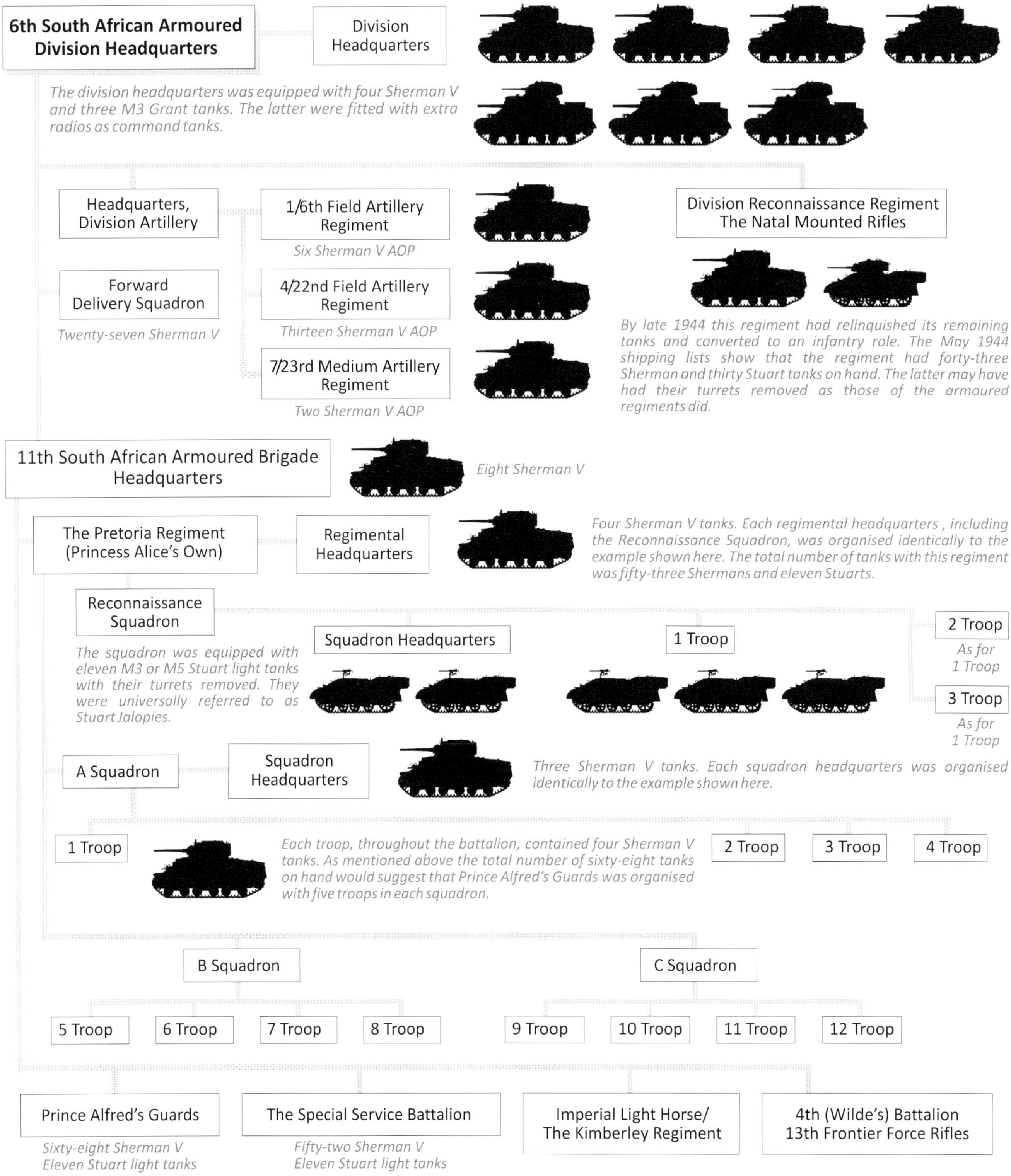


1. *Published in Scientia Militaria: South African Journal of Military Studies as* The Battle of Celleno - 10 June 1944 *and written by Evert Kleynhans, a senior lecturer with the Faculty of Military Science at Stellenbosch University, South Africa.*

## BRITISH AND COMMONWEALTH UNIT TANK STRENGTHS, 1943-1945

*All Allied units which took part in the fighting in Sicily and Italy came under the overall command of 15th Army Group. Regiments and battalions were supposed to report on their vehicle strength on a regular basis and these figures were used to compile returns for the whole army group. Not all these documents have survived the war but those that are available have been condensed into the table shown below. Not shown here are the numbers for specialist conversions based on the chassis of the Sherman tank such as bridgelayers, fascine carriers and Kangaroo personnel carriers, although they are mentioned in the accompanying text. It is true that the Duplex Drive (DD) wading tanks and Crab mine flails were designed with a specific purpose in mind but they were first and foremost gun tanks. I have included the figures for all British and Commonwealth armoured formations, as well as the Poles who were organised along British lines and used the same equipment, in order to present the reader with a more complete picture. I should mention that all the original reports, except that of 7 April 1945, employed American nomenclature which I have altered here to the British designations.*

| 9 July 1943 | Sherman IB | Sherman IC | Sherman II | Sherman IIA | Sherman III | Sherman V | Sherman VC | Sherman V | Sherman |
|---|---|---|---|---|---|---|---|---|---|
| | 105mm | Hybrid 17pdr | 75mm | 76mm | 75mm | 75mm | 17pdr | Crab | DD |
| 4th Armoured Brigade | | | | | 95 | | | | |
| 23rd Armoured Brigade | | | | | 76 | | | | |
| 1st Canadian Tank Brigade | | | | | | 58 | | | |

*The figures given above represent the number of tanks on hand for the invasion of Sicily and were in fact derived from individual unit records and convoy lists. None of the brigades deployed their full complement, with 1st Canadian Tank Brigade initially fielding a single regiment.*

| 2 October 1943 | Sherman IB | Sherman IC | Sherman II | Sherman IIA | Sherman III | Sherman V | Sherman VC | Sherman V | Sherman |
|---|---|---|---|---|---|---|---|---|---|
| | 105mm | Hybrid 17pdr | 75mm | 76mm | 75mm | 75mm | 17pdr | Crab | DD |
| 1st Armoured Division | | | | | | 18 | | | |
| 6th Armoured Division | | | | | 124 | | | | |
| 7th Armoured Division | | | | | 157 | | | | |
| 4th Armoured Brigade | | | 6 | | 199 | | | | |
| 23rd Armoured Brigade | | | | | 98 | | | | |
| 1st Canadian Armoured Brigade | | | | | | 145 | | | |
| 21st Tank Brigade | | | | | 5 | | | | |
| 25th Tank Brigade | | | | | 4 | | | | |
| Tank Replacement Group | | | 11 | | 23 | 46 | | | |
| Ordnance Stock, Training | | | 7 | | 333 | 248 | | | |
| and Maintenance units | | | | | | | | | |
| In Transit from UK | | | | | 285 | 71 | | | |
| In Transit from North America | | | 246 | | 72 | 117 | | | |

*1st Canadian Tank Brigade was renamed 1st Canadian Armoured Brigade at the conclusion of the fighting in Sicily.*

| 23 September 1944 | Sherman IB | Sherman IC | Sherman II | Sherman IIA | Sherman III | Sherman V | Sherman VC | Sherman V | Sherman |
|---|---|---|---|---|---|---|---|---|---|
| | 105mm | Hybrid 17pdr | 75mm | 76mm | 75mm | 75mm | 17pdr | Crab | DD |
| 1st Armoured Division | 17 | | 33 | 28 | 45 | 75 | | | |
| 6th Armoured Division | 9 | | | 2 | 215 | | | | |
| 5th Canadian Armoured Division | | | 46 | | | 174 | | | |
| 6th South African Armoured Division | | | | 14 | | 200 | | | |
| 7th Armoured Brigade | | | 105 | | | 56 | | | |
| 9th Armoured Brigade | *Reorganising as a specialist assault unit. Most of the original brigade returned to the UK in July and August 1944.* | | | | | | | | |
| 4th New Zealand Armoured Brigade | | | | | 163 | | | | |
| 1st Canadian Armoured Brigade | | | | | | 174 | | | |
| 2nd Polish Armoured Brigade | | | | | 164 | | | | |
| 3rd Hussars | | | | | 16 | | | | |
| 7th Hussars | | | | | 14 | 12 | | | |
| 21st Tank Brigade | | | 49 | | 4 | | | | |
| 25th Tank Brigade | | | 23 | | | | | | |
| Assault Regiment RAC/RE | | | | | 11 | 6 | | | |
| AOP Tanks, Royal Artillery | | | 8 | | 99 | 62 | | | |
| Tank Replacement Group | 4 | | 85 | 69 | 261 | 202 | | | |
| In Transit from UK | | 79 | | | | | | 25 | 90 |
| In Transit from North America | 91 | | | 82 | | | | | |

*Although it is not specifically stated in the report, the DD tanks were all based on Sherman III models.*

| 14 February 1945 | Sherman IB | Sherman IC | Sherman II | Sherman IIA | Sherman III | Sherman V | Sherman VC | Sherman V | Sherman |
|---|---|---|---|---|---|---|---|---|---|
| | 105mm | Hybrid 17pdr | 75mm | 76mm | 75mm | 75mm | 17pdr | Crab | DD |
| 6th Armoured Division | 12 | 4 | | 42 | 52 | | | | |
| 5th Canadian Armoured Division | 18 | 12 | 46 | | | 135 | | | |
| 6th South African Armoured Division | 18 | 7 | | 129 | | 76 | | | |
| 2nd Armoured Brigade | 18 | 11 | 4 | 114 | | 5 | | | |
| 7th Armoured Brigade | 19 | 12 | 47 | 61 | | 24 | | | |
| 9th Armoured Brigade | 4 | 4 | | | 43 | | | | 36 |
| 4th New Zealand Armoured Brigade | | 12 | | | 150 | | | | |
| 1st Canadian Armoured Brigade | 18 | 11 | | | | 152 | | | |
| 2nd Polish Armoured Brigade | | | | | 160 | | | | |
| 21st Tank Brigade | | | 30 | | 2 | | | | |
| Assault Regiment RAC/RE | | | | | 14 | 5 | | 5 | |
| AOP Tanks, Royal Artillery | | | 20 | | 82 | 87 | | | |
| 23rd Armoured Brigade | | | | | 37 | 34 | | | |
| *(In Greece)* | | | | | | | | | |
| Tank Replacement Group | 38 | 14 | 87 | 85 | 306 | 205 | | | 6 |
| Ordnance Stock, Training | 133 | 68 | 339 | 175 | 818 | 505 | 6 | 37 | 39 |
| and Maintenance units | | | | | | | | | |
| In Transit from UK | | 70 | | | | | | 5 | |
| In Transit from North America | | | | 36 | | | | | |

*The last strength report available to us today, that of 7 April 1945 which is shown below, was compiled at a time when the formations of 15th Army Group were being reinforced in preparation for a major offensive. Consequently many vehicles are not recorded here as being with units. Importantly for our study this includes the Sherman IC and Sherman VC 17pdr-armed tanks allocated to 6th South African Armoured Division which reported that four were allocated to each of the division's armoured regiments by early 1945. Similarly, it can be seen from the 7 April figures that large numbers of Sherman IB, Sherman III, Sherman V and the 17pdr conversions were recorded under the headings Tank Replacement Group, Ordnance Stock, Training & Maintenance or in were on their way from the UK. As these were the types operated by the New Zealanders and South Africans it could be expected that some were assigned to those units. By this time the Canadians had left Italy for the Netherlands but as they had taken all their armoured vehicles with them the Sherman II tanks that are seen with South African units in photographs taken in May 1945 probably came from the vehicles held by the Tank Replacement Group.*

| 7 April 1945 | Sherman IB 105mm | Sherman IC Hybrid 17pdr | Sherman II 75mm | Sherman IIA 76mm | Sherman III 75mm | Sherman V 75mm | Sherman VC 17pdr | Sherman V Crab | Sherman DD |
|---|---|---|---|---|---|---|---|---|---|
| 6th Armoured Division | 17 | 27 | 5 | 112 | 60 | | | | |
| 6th South African Armoured Division | 18 | 12 | | 126 | | 15 | | | |
| 2nd Armoured Brigade | 28 | 35 | | 98 | | 5 | | | |
| 7th Armoured Brigade | 9 | 12 | 6 | 32 | | 1 | | | |
| 9th Armoured Brigade | 4 | 5 | | | 16 | | | | 65 |
| *(without 14th/20th Hussars)* | | | | | | | | | |
| 4th New Zealand Armoured Brigade | 18 | | | | 122 | | 24 | | |
| 28th Armoured Brigade | 18 | 34 | | | 118 | | | | |
| 21st Tank Brigade | | | 4 | | | 6 | | | |
| 14th/20th Hussars | 4 | | | | | 18 | 6 | | |
| AOP Tanks, Royal Artillery | | | 24 | | 64 | 64 | | | |
| 25th Armoured Engineer Brigade | 3 | | | | 16 | 6 | | 11 | |
| 23rd Armoured Brigade | 1 | | | | | 69 | 2 | | |
| *(In Greece)* | | | | | | | | | |
| 311th Forward Delivery Squadron | 1 | | | | 50 | 22 | | | |
| *(In Greece)* | | | | | | | | | |
| Tank Replacement Group | 169 | 150 | 126 | 538 | 660 | 332 | 47 | 14 | 81 |
| Ordnance Stock, Training | 86 | 37 | 298 | 147 | 579 | 417 | 40 | 33 | |
| and Maintenance units | | | | | | | | | |
| In Transit from UK | | | | | | | 15 | | |
| In Transit from North America | | | | | | | | | |

*Raised in early March 1945, 28th Assault Squadron, New Zealand Engineers was specifically formed for the anticipated assault on the German positions along the Senio River. The squadron was made up of a headquarters and two troops with a total of twenty-one tanks. Initially it was to be equipped with specialist engineer vehicles based on Churchill and Valentine tanks but when these had failed to materialise by the first week of April, Sherman hulls were used. The fascine carrier shown here is a Sherman II, the so-called small hatch version, and may have been handed over from one of the British regiments which operated the type. The markings seen on this unit's vehicles varied considerably and the number on the hull side, 59 over A5, would seem to suggest some connection with the New Zealand Division's 5th Field Park Company. Note the extended towing coupling and the exhaust deflector.*

2

1

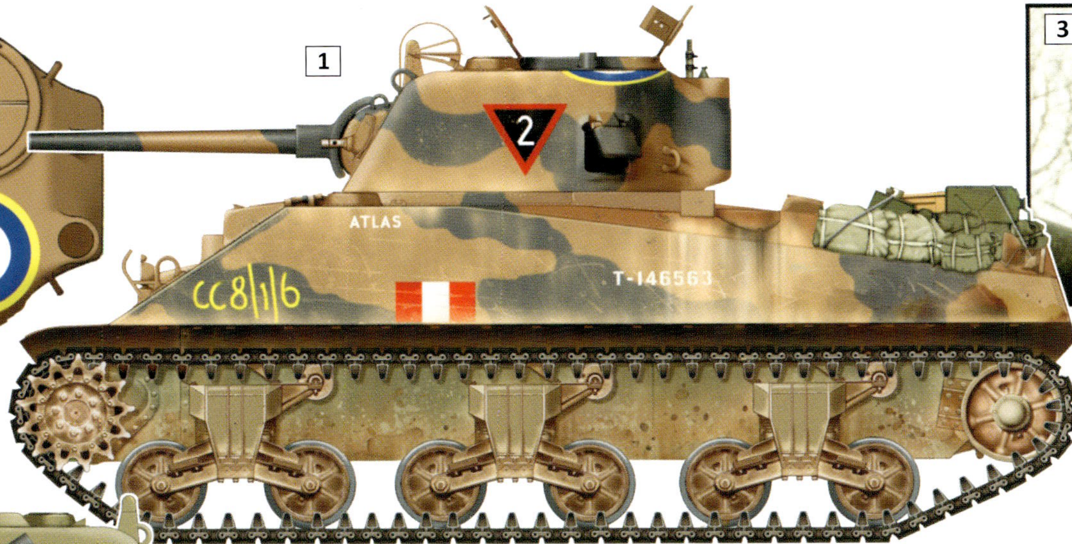


3

7

4

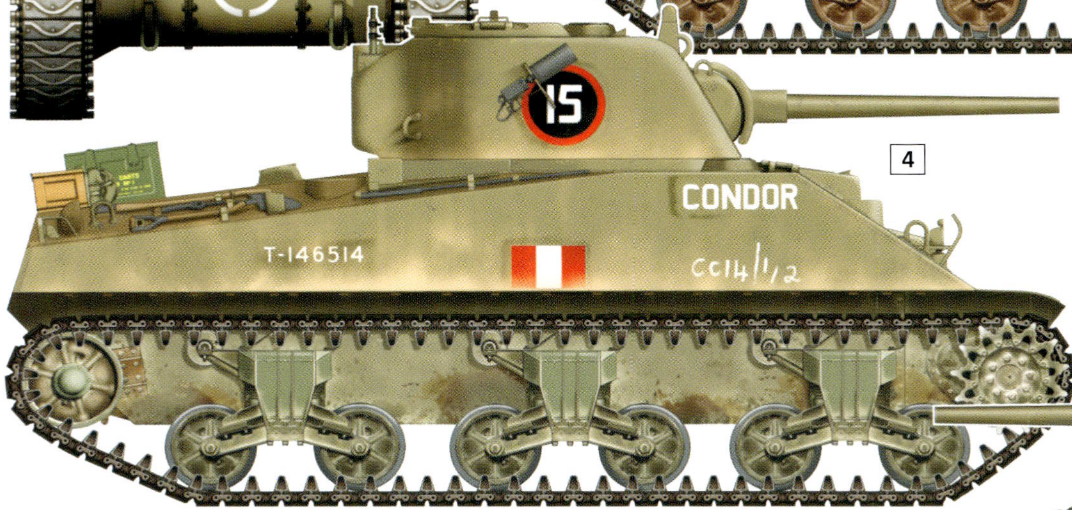


**1. Sherman V. A Squadron, 11th Armoured Regiment (The Ontario Regiment), 1st Canadian Armoured Brigade. Sicily, August 1943.** In May 1943, in preparation for the landings in Sicily, the regiments of this brigade were ordered to camouflage their tanks using Standard Camouflage Colour (SCC) 4 Stone and SCC 14 Black and to apply the RAF roundel aerial recognition device to the upper surface of the turret (2). Note also the red-white-red RAC flash on the hull side introduced in mid-1942 as a general recognition sign and used extensively in Italy. The Swabey sight (3) fitted to the turret front was often seen in Italy and was also fitted to British Shermans operating in Burma.

6

5

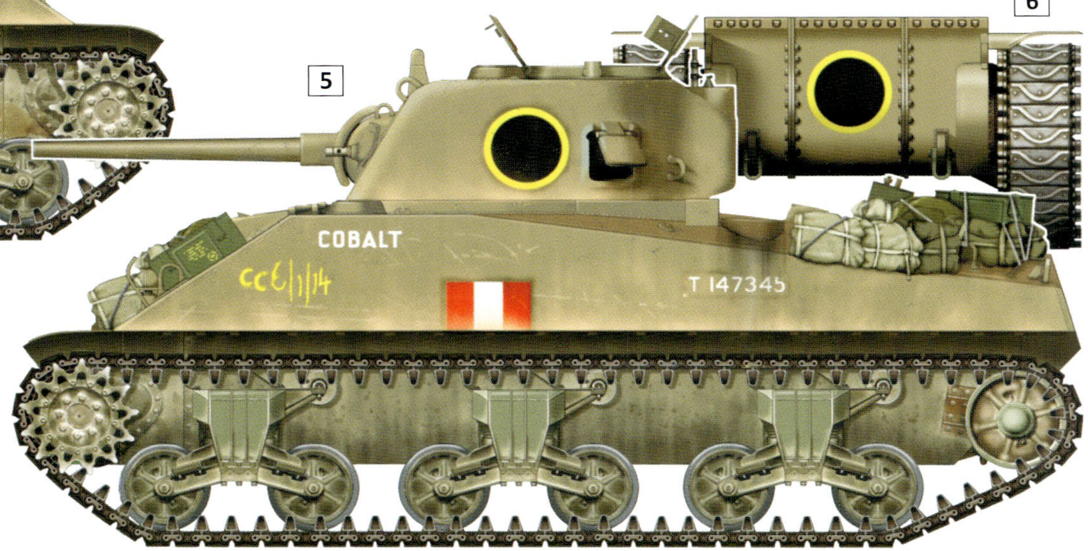


**4. Sherman V. C Squadron, 11th Armoured Regiment (The Ontario Regiment), 1st Canadian Armoured Brigade. Sicily, August 1943.** Available photographs show clearly that not all the brigade's tanks were camouflaged and many may have retained their base coat of US Army Olive Drab. Prior to embarkation the markings were covered in paper which was held in place by varnish. When the paper was removed the residue of the varnish could be seen for some time as a darker surround.

**5. Sherman V. C Squadron, 12th Armoured Regiment (The Three Rivers Regiment), 1st Canadian Armoured Brigade. Sicily, August 1943.** All the tanks of this regiment were marked with the prominent squadron sign shown here which was also repeated on the hull front (6). This seems to have been replaced throughout the brigade by a white broken circle (7) before the first landings on the Italian mainland.

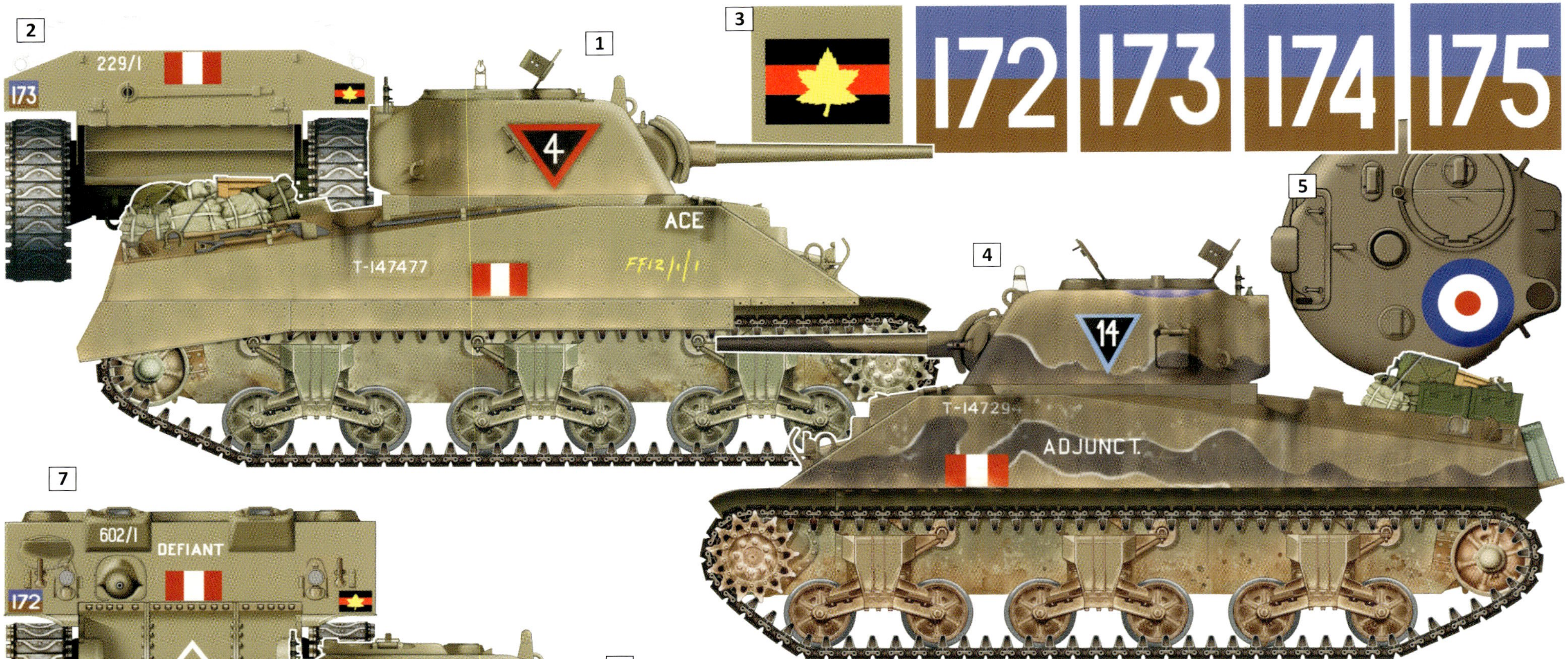


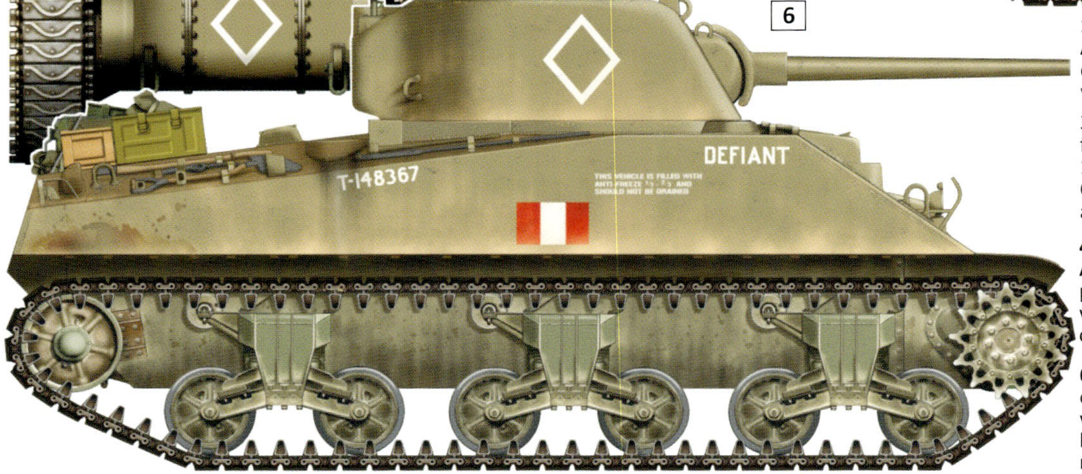


**1. Sherman V. A Squadron, 11th Armoured Regiment (The Ontario Regiment), 1st Canadian Armoured Brigade. Sicily, August 1943.** Prior to the invasion of Sicily the brigade was ordered to display the formation badge, unit serial number and mobilisation number, here 229/1 (2). The latter was unique to each Canadian unit and the 1 signified overseas service.

3. The unit formation badge of 1st Canadian Armoured Brigade with the unit serial numbers of, from left to right, the brigade headquarters, 11th Armoured Regiment (The Ontario Regiment), 12th Armoured Regiment (The Three Rivers Regiment) and the 14th Armoured Regiment (The Calgary Regiment). Each number was displayed on an Arm of Service square of blue over brown although the exact shades are the subject of some debate.

**4. Sherman V. A Squadron, 14th Armoured Regiment (The Calgary Regiment), 1st Canadian Armoured Brigade. Italy, September 1943.** Although somewhat speculative there is some possibility that this tank was camouflaged with SSC 2 Brown and SCC 1A Dark Brown, a scheme which was in use with units in the UK. The lighter dividing line is more difficult to explain but is clearly visible in the photograph of this tank. Note that the RAF roundel lacks the yellow outer ring.

**6. Sherman V. Brigade headquarters, 1st Canadian Armoured Brigade. Italy, early 1944.** The tanks of the brigade headquarters were identified by the diamond-shaped tactical sign depicted in white which denoted an unbrigaded unit. Photographs of other headquarters tanks suggest that they may have all carried names beginning with the letter D. The orders specifically stated that the mobilisation number was to be applied to the rear hull but here it is displayed at the front (7).

4

2

1

**1. Sherman V. Divisional headquarters, 5th Canadian Armoured Division. Italy, early 1944.** This tank was the personal mount of the division's commander, Major-General Bert Hoffmeister, and is painted in US Army Olive Drab, which most of the division's tanks were at this time. The tactical sign identifies a headquarters unit and the number nine may have been selected at random. The tank's name and the unit insignia were repeated on the hull front (2). Most Canadian units removed the anti-aircraft machine gun from the cupola but is is just visible in the photograph on which this illustration is based.

3

**3. Sherman V. C Squadron, 12th Armoured Regiment (The Three Rivers Regiment), 1st Canadian Armoured Brigade. Italy, December 1943.** The appearance of the brigade's tanks changed little throughout the last months of 1943 and this vehicle is remarkable only in repeating the census number, T 147291, on the hull side. Note the ammunition box welded to the turret rear.

4. The formation badge of 5th Canadian Armoured Division followed by, from left to right, division headquarters, 5th Armoured Brigade headquarters, 2nd Armoured Regiment (Lord Strathcona's Horse), 5th Armoured Regiment (8th Princess Louise's (New Brunswick) Hussars) and 9th Armoured Regiment (The British Columbia Dragoons). The black Arm of Service square was used to identify divisional headquarters troops while the red square denoted an armoured brigade attached to a division. The style of numbering could vary and all those shown here were taken from actual examples.

5

6

**1. Sherman V. B Squadron, 5th Armoured Regiment (8th Princess Louise's (New Brunswick) Hussars), 5th Canadian Armoured Brigade. Italy, December 1944.** The Light Mud and Black camouflage pattern, applied to this tank in accordance with the April 1943 instructions, would suggest that it was taken over from a British regiment. In addition, it is fitted with the Crusader-style blanket box and sunshield rail which are rarely seen on Canadian tanks. Note the early gun mantlet, the protective cover for the turret machine gun and that the turret pistol port has been welded shut. The B Squadron tactical sign, enclosing the troop number, is repeated on the blanket box.

**1. Sherman II. C Squadron, 3rd Armoured Reconnaissance Regiment (The Governor General's Horse Guards), 5th Canadian Armoured Division. Italy, spring 1945.** The division had a significant number of Sherman II tanks on hand and most seem to have been concentrated in the reconnaissance regiment. Note the simpler vane sight on the turret front and thickened armour of the hull side. The latter is examined further in the Technical Details and Modifications section of this book. The name Cobourg III (2) is clearly visible in our photograph as is the pintle-mounted machine gun.

1

**3. Sherman V. 5th Canadian Armoured Brigade. Italy, spring 1945.** Photographed as it arrived in Marseille in March 1945 after leaving Italy, this vehicle has been newly repainted in SCC 15 as most, but not all, the brigade's tanks were. I can offer no explanation as to the meaning of the tactical sign on the turret side. It is not a circle but a letter O as shown here and is quite clear in our photograph. Note the non-standard blanket box, the cover for the smoke discharger triggers and the remains of the Swabey sight on the turret.

3

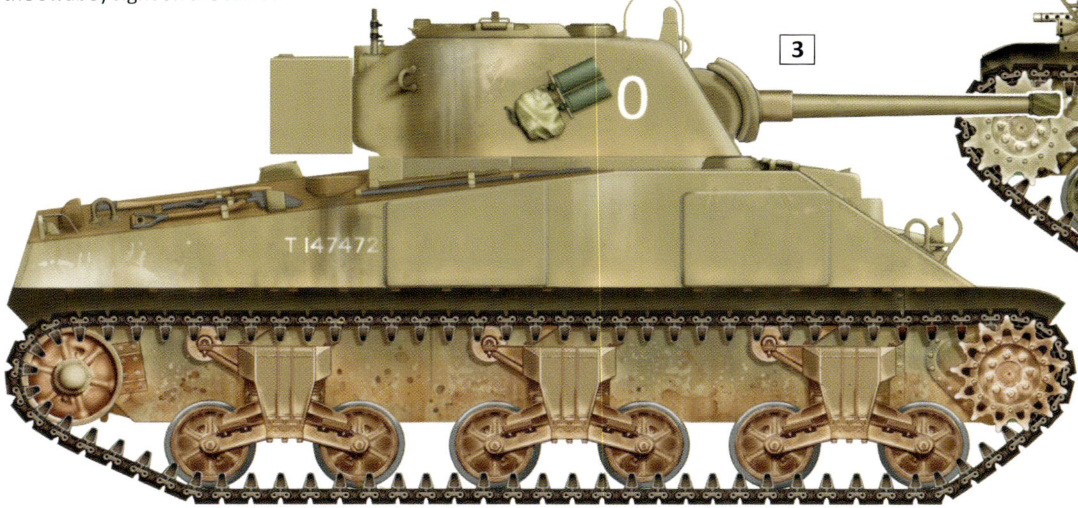


2

6

**4. Sherman II. B Squadron, 3rd Armoured Reconnaissance Regiment (The Governor General's Horse Guards), 5th Canadian Armoured Division. Italy, spring 1945.** Photographed as the regiment was preparing to leave Italy for the Netherlands, this tank is a somewhat earlier production version than the Sherman II shown at the top of the page, lacking the metal strip along the lower edge of the hull to hold the sandshields and it also has the welded-on applique armour. The Allied recognition (5) star on the turret roof had almost completely replaced the RAF roundel by this time. The green over blue Arm of Service square with the unit serial number 45 (6) displayed on the hull front (7) identified a reconnaissance regiment attached to an armoured division. The name Bunyon is speculative as it is not clear in our photograph but is known to have been used by this regiment.

5

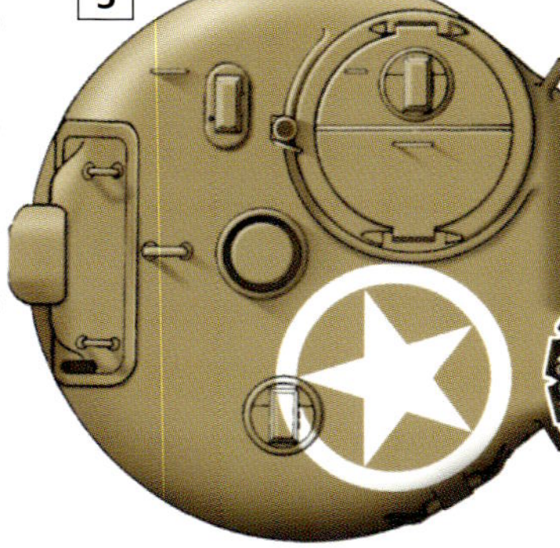

4

7

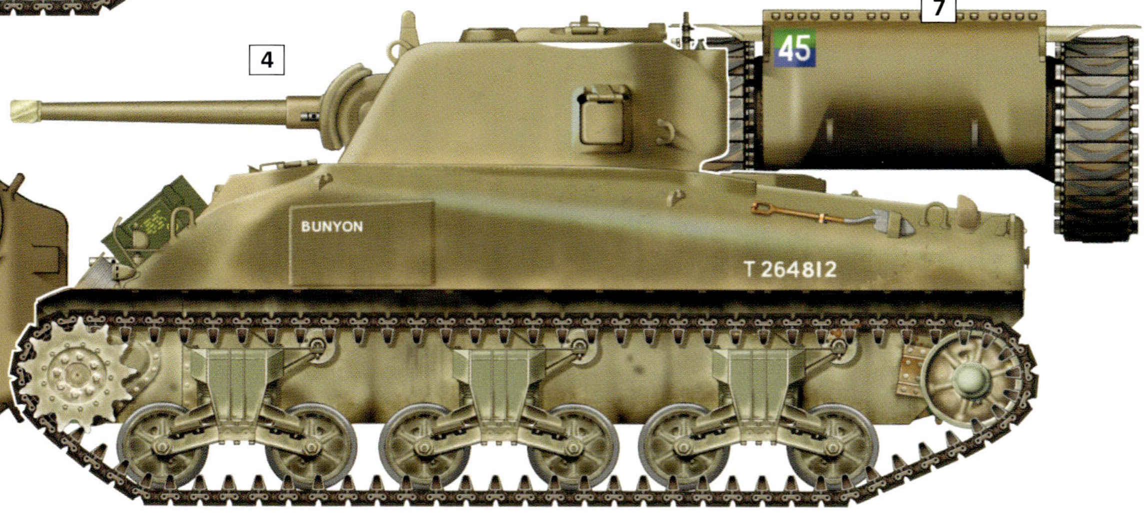

2

Above: Unit serial numbers and Arm of Service squares identifying elements of 4th New Zealand Armoured Brigade. From left to right, brigade headquarters, 18th Armoured Regiment, 19th Armoured Regiment and 20th Armoured Regiment. The application, size, style and the rendering of the silver fern formation badge varied considerably and the examples shown here are all based on contemporary photographs.

1

**1. Sherman III. C Squadron, 19th Armoured Regiment, 4th New Zealand Armoured Brigade. Italy, spring 1944.** Before leaving Egypt for Italy the tanks were camouflaged in Light Mud and Black, following the April 1943 instructions. Most tanks also carried the RAF roundel on the engine deck (2) although the example shown here lacks the outer yellow ring. The squadron tactical sign, which has a dark inner circle, encloses the troop number while the tank's individual number is displayed on the hull side. These large numbers were a common feature of the brigade's tanks, especially in the early stages of the campaign.

**3. Sherman III. Divisional Protection Troop, 2nd New Zealand Division. Italy, spring 1944.** These tanks were camouflaged in an identical fashion to those of the 4th Armoured Brigade but carried no insignia other than the unit serial number 50 on a black square (4) which denoted the division headquarters. Note the two Crusader-style blanket boxes on the hull rear plate.

**5. Sherman III. 18th Armoured Regiment, 4th New Zealand Armoured Brigade. Italy, summer 1943.** This tank is painted in the short-lived camouflage scheme of green with a brown disruptive employed by this regiment during May and June 1944. The circular markings on the hull side were used to bore-sight the guns of the other tanks of the regiment.

3

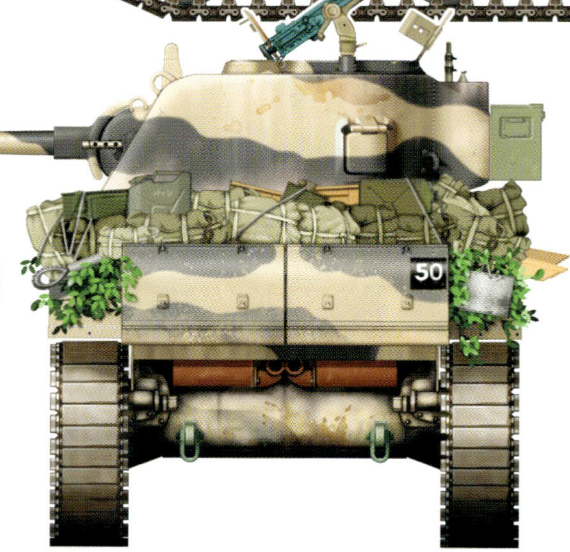


4

5

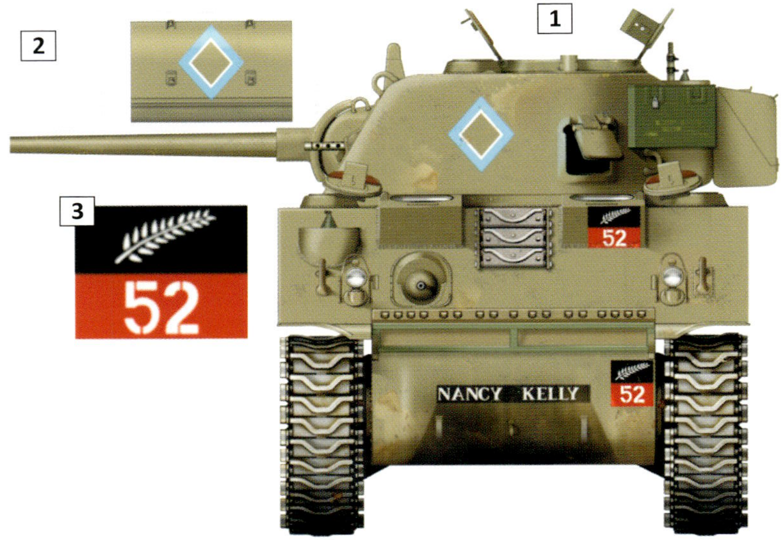


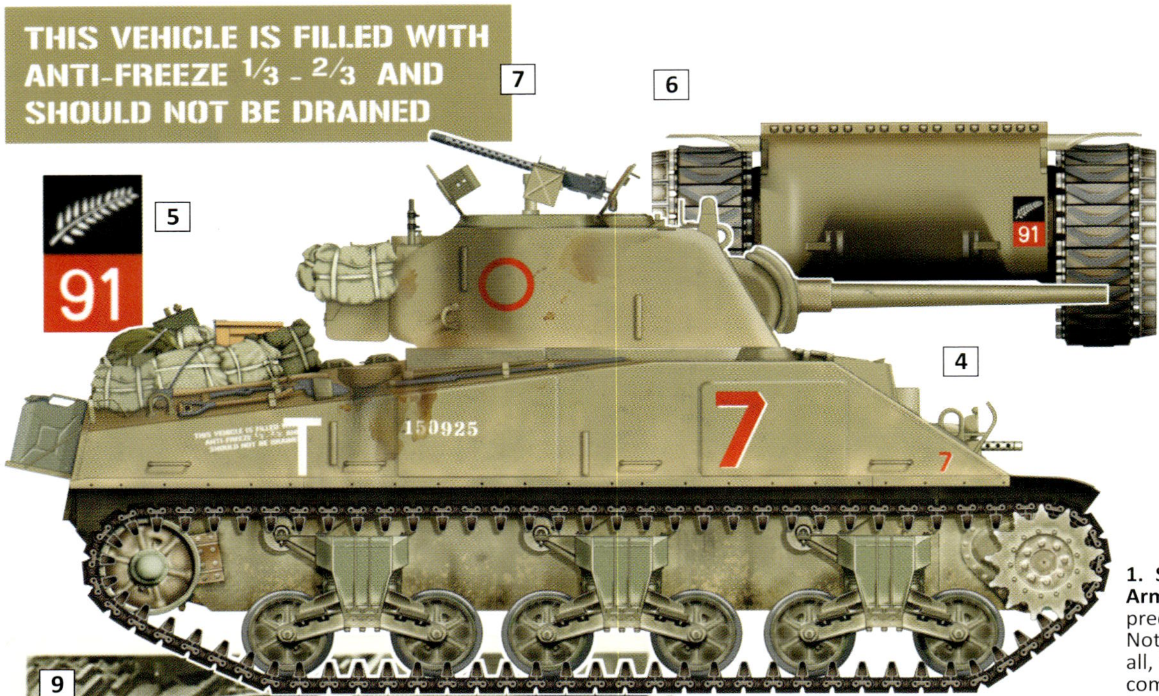


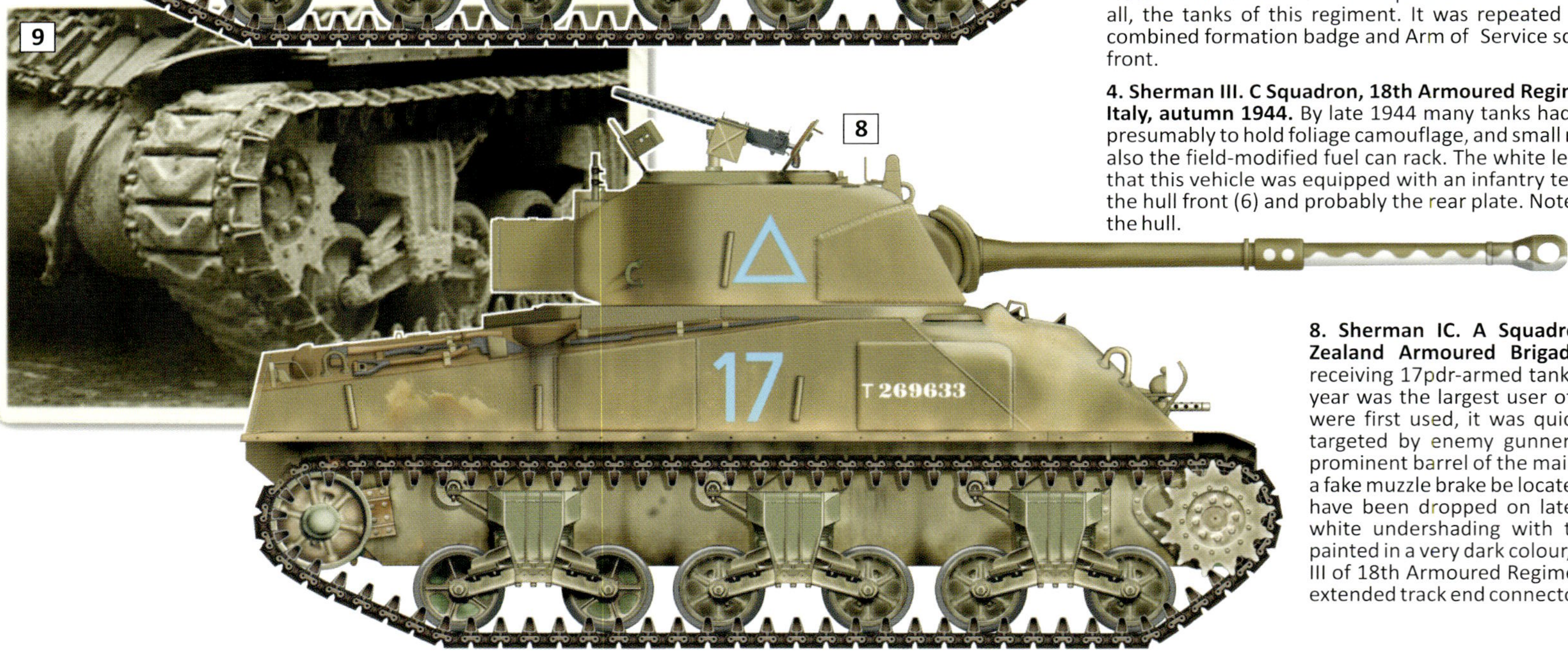


**1. Sherman III. Headquarters Squadron, 20th Armoured Regiment, 4th New Zealand Armoured Brigade. Italy, summer 1944.** By this time the Olive Drab camouflage was predominant although the Light Mud and Black scheme could be seen until well into 1945. Note the white centre of the squadron tactical marking which was common to most, but not all, the tanks of this regiment. It was repeated on the blanket box (2) and unusually the combined formation badge and Arm of Service square (3) has been applied twice on the hull front.

**4. Sherman III. C Squadron, 18th Armoured Regiment, 4th New Zealand Armoured Brigade. Italy, autumn 1944.** By late 1944 many tanks had tubes welded to the hull and turret sides, presumably to hold foliage camouflage, and small metal handles welded to the hull sides. Note also the field-modified fuel can rack. The white letter T on the hull towards the rear signified that this vehicle was equipped with an infantry telephone. Unit markings (5) were applied to the hull front (6) and probably the rear plate. Note the anti-freeze warning (7) stencilled onto the hull.

**8. Sherman IC. A Squadron, 20th Armoured Regiment, 4th New Zealand Armoured Brigade. Italy, late 1944.** The brigade began receiving 17pdr-armed tanks in late 1944 and by April of the following year was the largest user of the type. In France, where these vehicles were first used, it was quickly discovered that they were specifically targeted by enemy gunners and efforts were made to disguise the prominent barrel of the main gun. The official instructions directed that a fake muzzle brake be located halfway along the barrel but this seems to have been dropped on later modifications. Variations existed to the white undershading with the top portion of the barrel sometimes painted in a very dark colour, probably black. This tank, and the Sherman III of 18th Armoured Regiment shown above (4), were both fitted with extended track end connectors (9).

**1. Sherman VC. C Squadron, 19th Armoured Regiment, 4th New Zealand Armoured Brigade. Italy, early 1945.** As numbers of 17pdr conversions based on the Sherman V became available they replaced many of the Sherman IC models. Note the very elaborate paintwork on the main gun barrel which incorporates a lighter green shade. These later arrivals were also fitted with the specially designed stowage box fitted to the hull rear (2) and most had exhaust deflectors and smoke emitters. These latter details are shown in photographs in the main text. Note the handles on the hull side and the rectangular blanket box which was a feature of many of the New Zealand Fireflies. The combination formation badge and Arm of Service square (3) was probably repeated on the hull front.

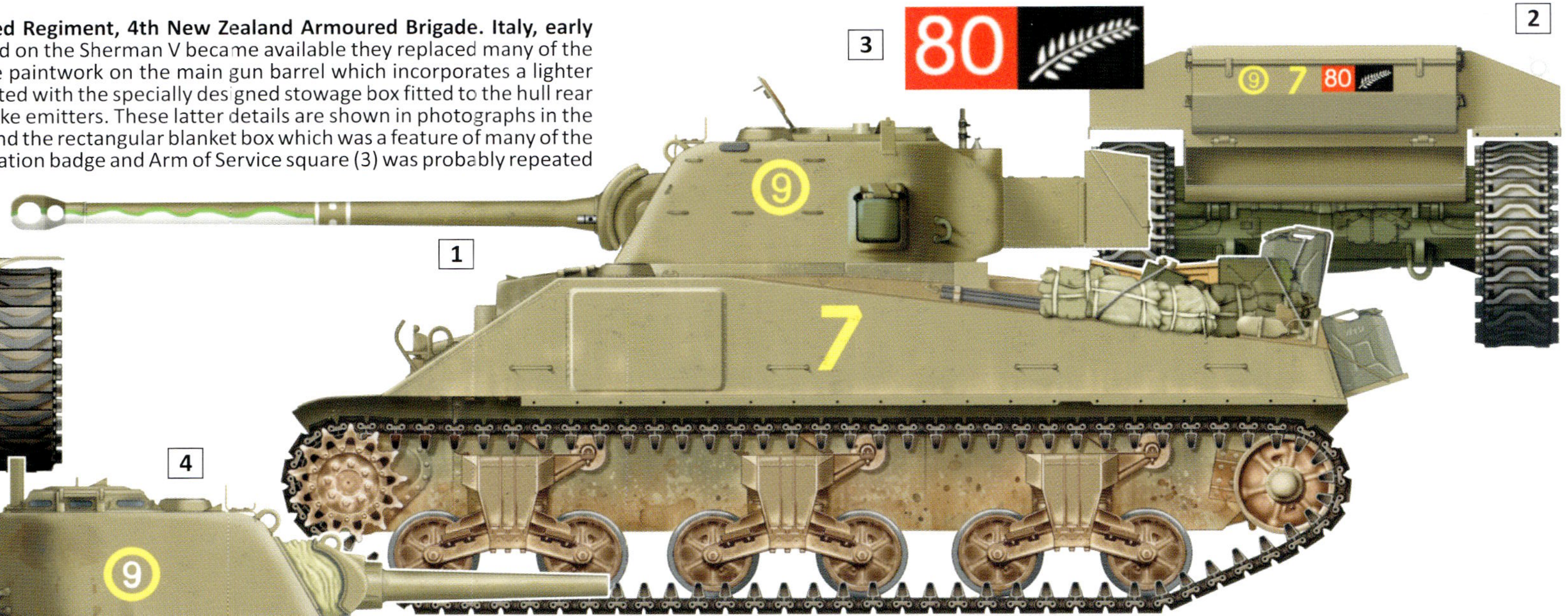


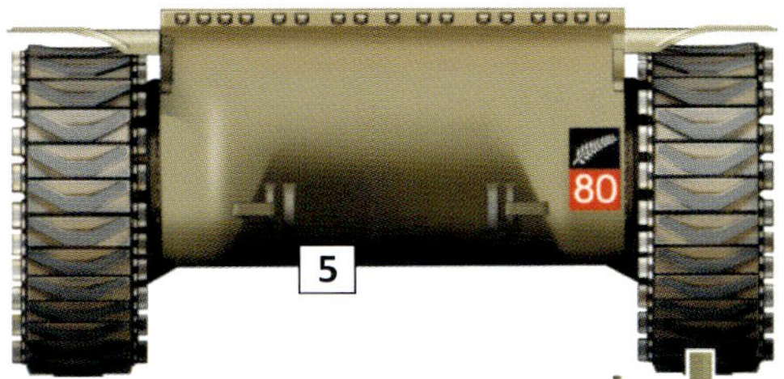

**4. Sherman IB. C Squadron, 19th Armoured Regiment, 4th New Zealand Armoured Brigade. Italy, spring 1945.** The brigade received just eighteen of these 105mm-armed tanks and they were originally allocated on a basis of two per squadron in each regiment. Although this arrangement was later changed they seem to have retained their original markings. Note the rectangular blanket box which appears to have largely replaced the Crusader-style version, at least on the late arrivals. The placement and style of the unit markings (5) are quite clear in the photographs used to create this illustration.

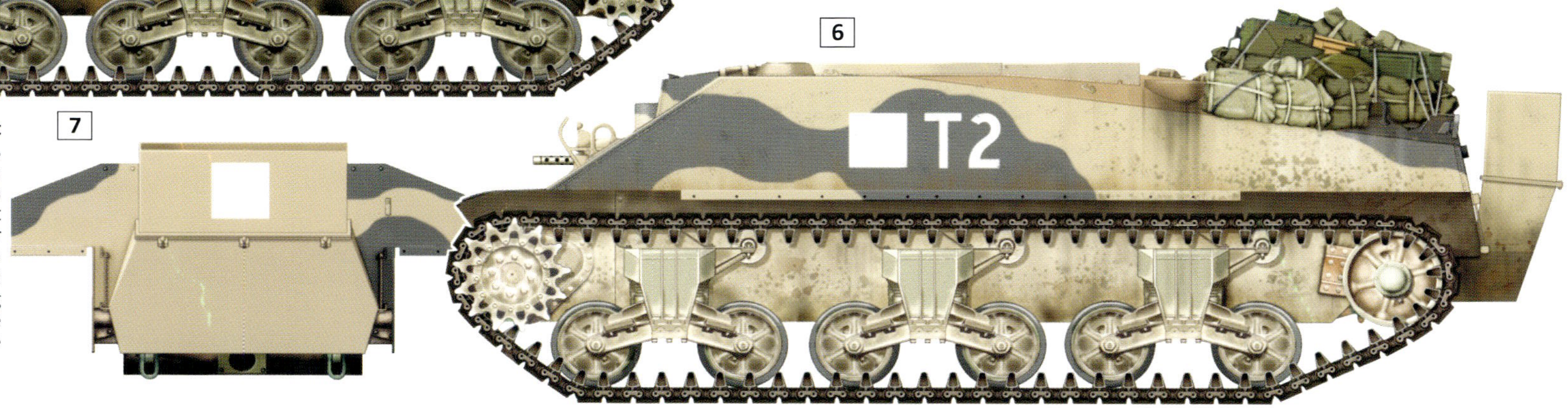


**6. Sherman III Kangaroo. 28th Assault Squadron, New Zealand Engineers. Italy, spring 1945.** The New Zealand Division adapted the mechanically reliable Sherman tank to undertake a number of specialist roles including the personnel carrier depicted here. This simple conversion merely required that the turret be removed but this vehicle has also had a deep wading stack fitted to the engine exhaust (7). The presence of a disruptive camouflage pattern is quite clear in our photograph.

2

1

3

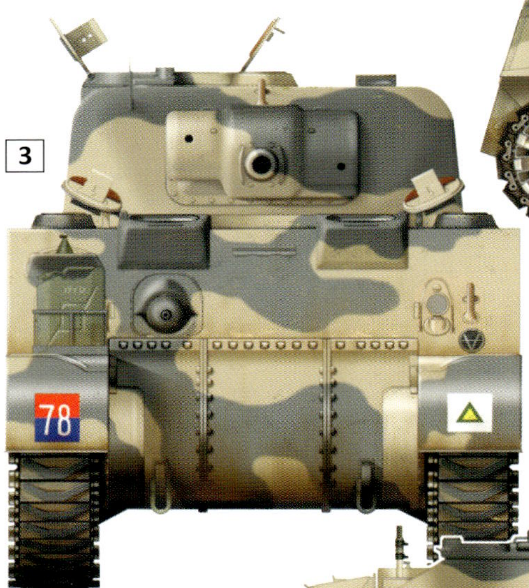


4

5

6

8

9

**1. Sherman V. C Squadron, The Pretoria Regiment (Princess Alice's Own), 11th South African Armoured Brigade. Italy, spring 1944.** Like the New Zealand regiments the South Africans were equipped in Egypt and the Shermans were camouflage in Light Mud and Black before they departed for Italy. This tank is unusual in that the Olive Drab base coat has been incorporated into the scheme. Note the Swabey sight and the Crusader blanket box. The right side of the turret is not actually visible in our photograph and I have assumed that the April 1943 instructions regarding the disruptive camouflage pattern were followed. The left side of the turret is shown above (2).

**3. Sherman V Armoured Observation Post (AOP). 7/23 Medium Artillery Regiment, 6th South African Armoured Division. Italy, spring 1944.** This tank features an almost textbook application of the Light Mud and Black camouflage, merely lacking the white undershading on the hull front. It is likely that most of the 6th Armoured Division's Sherman tanks were fitted with sandshields and some can still be seen in 1945 with the division's formation badge (4) carried in the position shown here. The unit serial number 78 (5) denotes the junior regiment of the artillery group. As the regiment had just two Sherman tanks on hand it would seem reasonable to assume that one was allocated to the commander and the appropriate tactical sign (6) may have been painted onto the turret side.

7

**7. Sherman V AOP. 4/22 Field Artillery Regiment, 6th South African Armoured Division. Italy, spring 1944.** This tank is also fitted with the Swabey sight and is painted in the Light Mud and Black camouflage scheme. The artillery tactical sign on the turret side denotes a battery commander (8). Other AOP tanks of this regiment carried the Arm of Service square with unit serial number 76 (9) on the hull rear plate identifying the second senior regiment of the division's artillery group but it is not visible in the photograph of this vehicle. Note the rail for the sunshield on the hull side which was fitted to many of the division's tanks in Egypt.

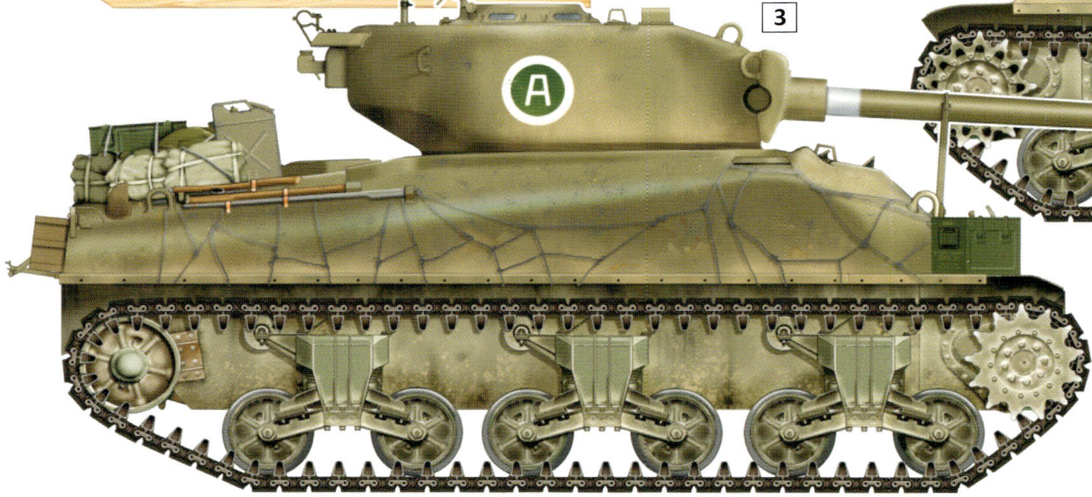

**1. Sherman V AOP. 4/22 Field Artillery Regiment, 6th South African Armoured Division. Italy, late 1944.** This tank may very well be one of the replacements for the Sherman V AOP named Pat shown on the previous page. Note the blanket box, the stowage bin on the hull side and the additional armour plate welded to the hull in front of the driver's hood. The division's formation sign was carried on the hull front (2) and the artillery tactical sign identifies the battery's second in command. The markings on the hull side are a variation of the bore-sight calibrations.

**3. Sherman IIA. C Squadron, Special Service Battalion, 11th South African Armoured Brigade. Italy, spring 1945.** The identification of this tank with the brigade's junior armoured regiment is based on a direction indicator (4) stowed on the hull front, visible in our photograph, which clearly shows the unit serial number 53. The C Squadron tactical insignia on both the wooden sign and the turret is far too light to have been depicted in the green that was authorised for this formation and it may be that at this time the regiment was considered an unbrigaded unit. The Special Service Battalion later reverted to green tactical signs as can be seen on the following page. Note the wire attached to the hull to hold foliage camouflage. Other unit serial numbers and Arm of Service markings of 6th South African Armoured Division are shown above (5) with, from left to right, the division headquarters, The Natal Mounted Rifles (the division's reconnaissance regiment), 11th Armoured Brigade headquarters, Prince Alfred's Guard and The Pretoria Regiment.

**6. Sherman VC. C Squadron, The Pretoria Regiment (Princes Alice's Own), 11th South African Armoured Brigade. Italy, spring 1945.** South African units received a small number of 17pdr-armed tanks in late 1944 as a trial but significant numbers were not available until after mid-April 1945. This tank has a rather elaborate camouflage scheme applied to the barrel of the main gun, as laid down in official instructions, and the squadron sign is repeated on the turret rear extension. The letters IA (7)may indicate the individual tank and the troop number.

4

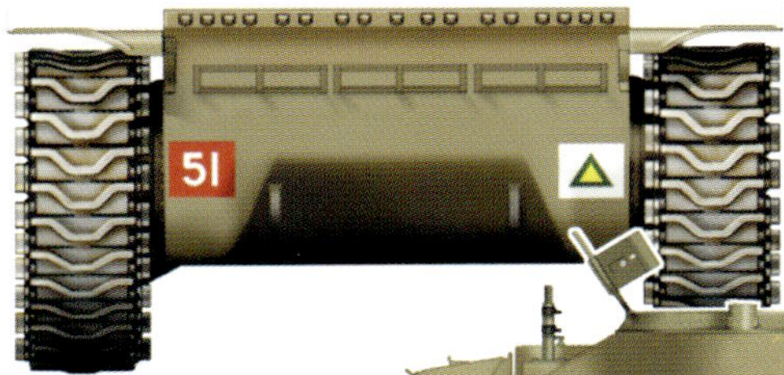


3

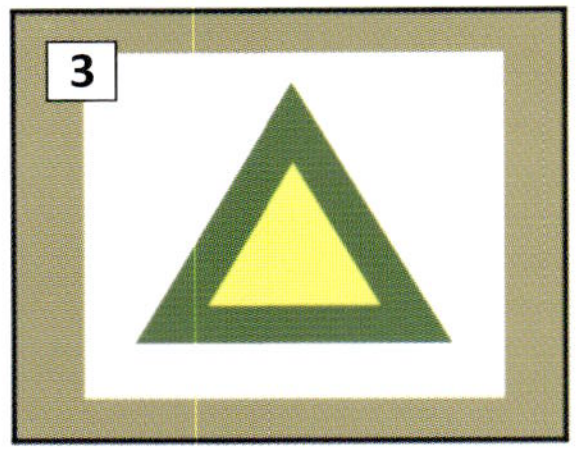

1

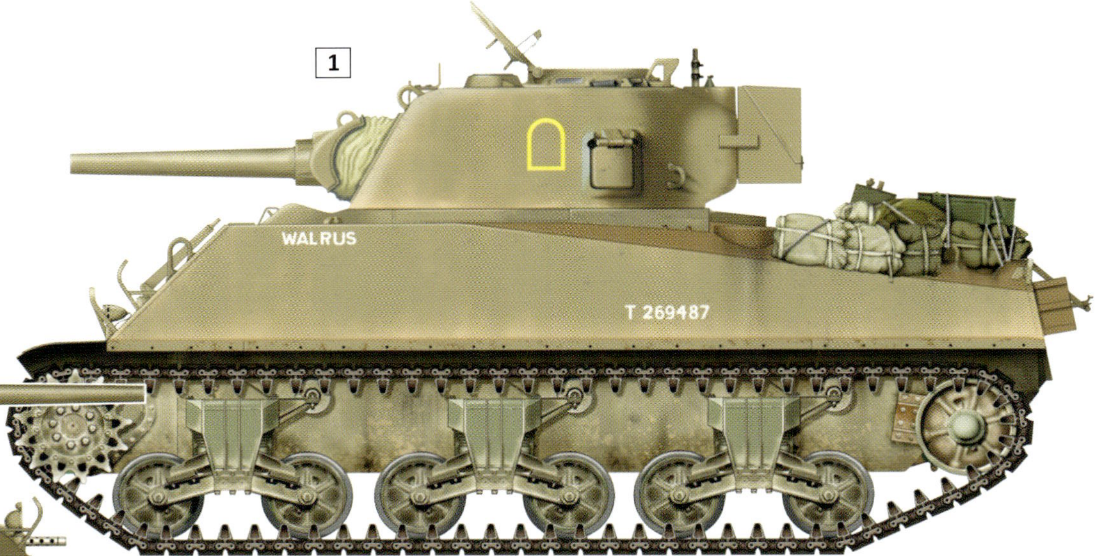


2

**1. Sherman V. D Squadron, The Pretoria Regiment (Princes Alice's Own), 11th South African Armoured Brigade. Italy, spring 1945.** Each squadron of this regiment was initially issued three of these 105mm-armed tanks and although all were later grouped together under the direct command of regimental headquarters they seem to have retained their original markings. Note the name Walrus, painted in white, and the rarely-seen D Squadron tactical sign.

**2. Sherman II. B Squadron, Prince Alfred's Guard, 11th South African Armoured Brigade. Italy, spring 1945.** Photographed during the victory parade at Monza in May 1945, this tank was probably handed over by one of the British regiments as they left Italy as there is no record of this type being allocated to the South Africans. The division's formation badge (3) and the unit serial number which identifies the regiment were both carried on the hull front (4) as shown here. Note the bracket welded to the cast transmission cover. These are still in evidence on tanks displayed at a number of museums throughout South Africa today.

5

6

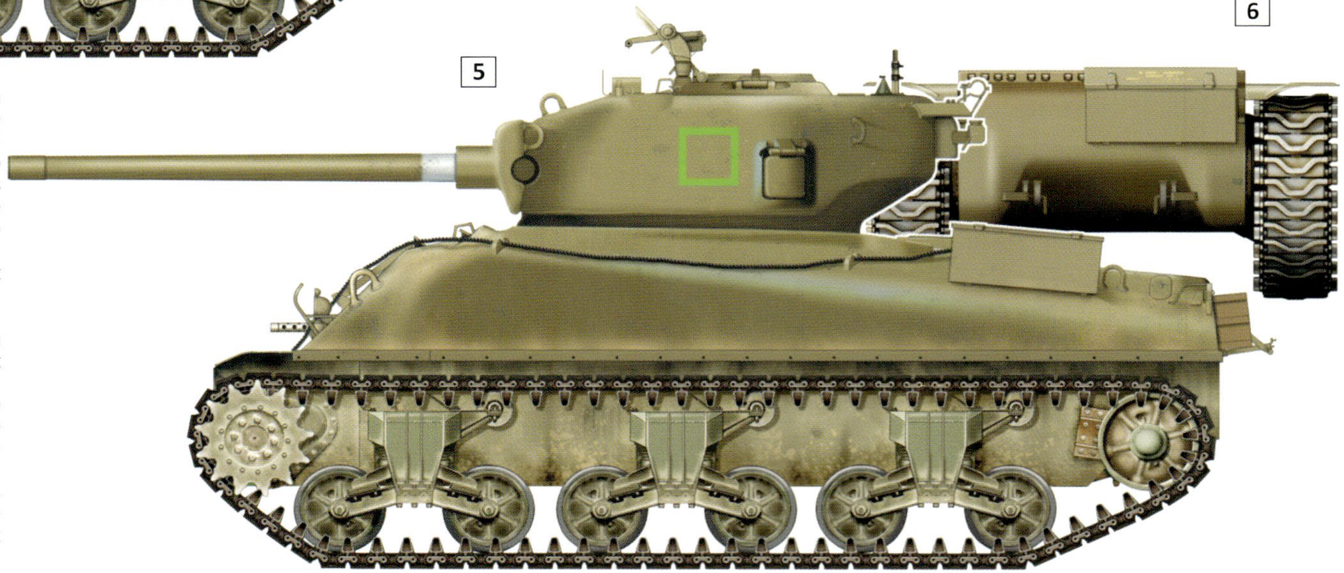

**5. Sherman IIA. B Squadron, Special Service Battalion, 11th South African Armoured Brigade. Italy, spring 1945.** Also photographed at Monza, this tank carries the squadron tactical sign in light green. Originally this marking was a much darker shade which would have shown up adequately on a background of Light Mud but would have been too similar to US Army Olive Drab or SCC 15. The possibility of the temporary use of white is explored on the previous page. I have based the green shade here on that shown in a post-war regimental veterans association publication applied to a Sherman tank. Note the stowage box on the hull side. Many of the regiment's tanks photographed at this time also had a large box fixed to the hull front (6).

# SHERMAN III

## 4TH NEW ZEALAND ARMOURED BRIGADE

1/35 SCALE

GARY BOXALL

*Gary's replica, which depicts a Sherman III of 20th New Zealand Armoured Regiment, is based on the 1/35 scale Asuka Sherman III Mid Production, above, and completed with markings from Star Decals, shown at right. These waterslide transfers are offered in both 1/35 and 1/72 scale are also shown on page 48. A photograph of the actual vehicle, which was disabled during the fighting around Cassino in March 1944, is reproduced on the previous page.*

*A view of the completed model. Gary added counterbalance springs to the hull hatches along with brass chains for the various filler caps. Photo-etched brass details from a number of manufacturers provided new light guards, tool straps, sunshield rails and the Petrol Oil Water (POW) can rack on the front fender. The machine gun barrels are from Aber and the 75mm gun barrel is a turned aluminium product from RB Models. Once assemble the model was sprayed with Vallejo Khaki 988 and the pattern masked off. The dark disruptive camouflage was created using LifeColor RAL 7021 PanzerGrau and a clear gloss applied to the segments where the markings would be applied. These are available from Star Decals and this set is shown on page 48. The stowage bins were both scratch-built from plastic sheet. Note the muddy boot prints on the engine deck.*

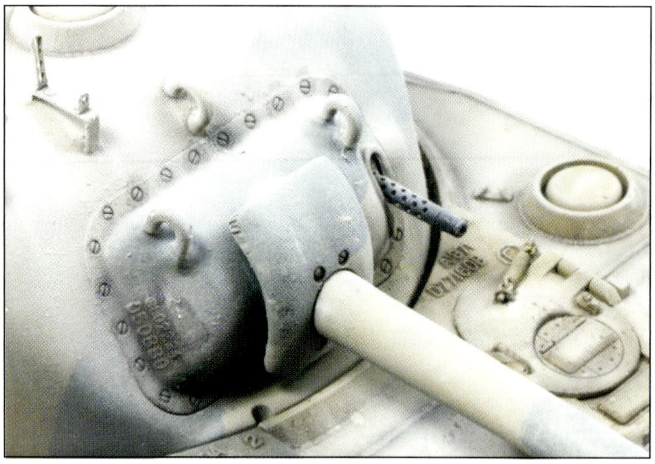

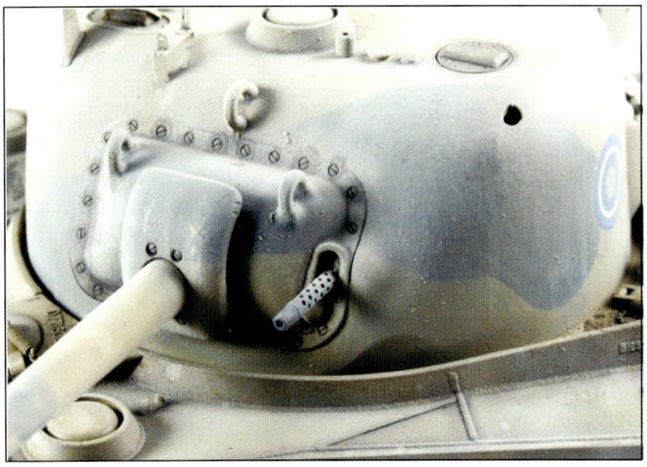

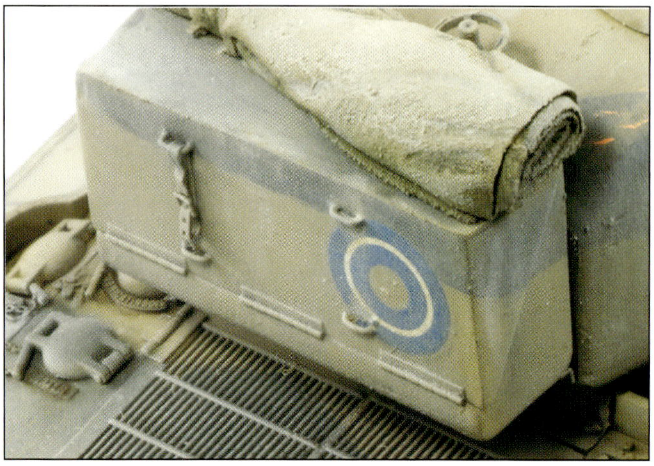

*With the markings in place the entire vehicle was given a matt coat and once dry, filters and washes from AMMO by Mig Jimenez and AK Interactive. Tools were painted using a matt flesh colour for the base and then adding some raw sienna oil paint which was left for about an hour before a soft brush, lightly moistened with thinners, was dragged along the surface to create a wood grain effect. Other details such as machine gun barrels were also picked out at this stage with a fine brush. Rain and Dust Effects from AK were also used to add that well-used look. The tracks were first painted using Vallejo Black primer with the rubber pads rendered in LifeColor Tyre Black with rust-coloured end connectors. Further pigments blended the tracks in with the rest of the vehicle.*

*Built-up mud was added to the lower hull sides by mixing different pigments and water into a slurry which was applied with an old brush to stipple the mixture to the relevant areas. Further pigments were added until the desired effect was achieved. Photographs of the actual tank, including that reproduced on page 27, seem to show that the vehicle was not so much worn or damaged but rather dirty and dusty and that is the look Gary achieved.*

# SHERMAN V

## 1ST CANADIAN ARMOURED BRIGADE

## 1/35 SCALE

JORDAN BAKER

Jordan's model depicts a Sherman V from A Squadron, The Calgary Regiment of 1st Canadian Armoured Brigade. This tank, named Arrant, is pictured at right and is also shown in the photograph on page 4. The finished project was based on the Dragon 1/35 scale M4A4 Sherman kit which was originally released in 1955 and the suspension from Dragon's Sherman III kit

*The basic model fully assembled, except for the tracks and headlights, showing the brass details including the radio antennae, manufacturer's cast mark, headlight brushguards and machine gun. Note the RAC helmet under the opened driver's hatch which can be seen in the photograph of the actual vehicle reproduced on the previous page.*

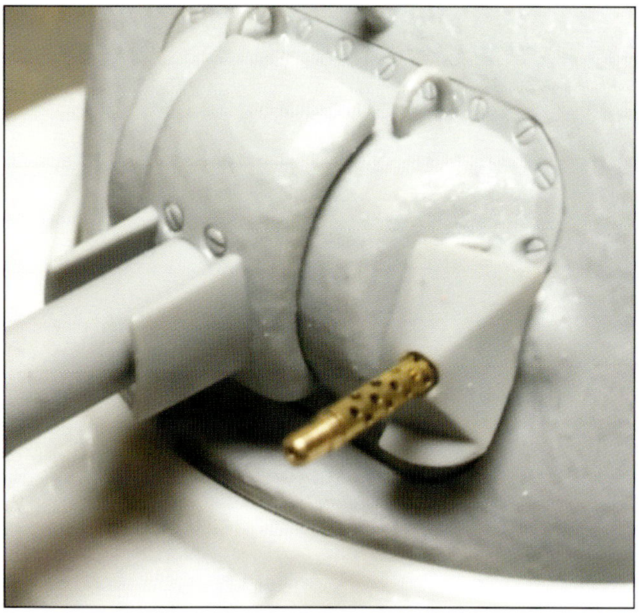

*The partly assembled model was painted with Tamiya acrylic colours to simulate the Standard Camouflage Colour (SCC) 4 Stone and SCC 14 Black scheme that was commonly seen on the Canadian tanks which took part in the landings in Sicily and the early battles on the Italian mainland.*

*The tank's markings were, for the most part, not commercially available and Jordan created the accurate insignia from parts of various transfers. The next step was to weather the hull and turret with oil washes and pigments.*

*The finished model showing the extensive weathering of the tracks and hull. The images below shown the driver and the hatch interior with its periscope, the turret pistol port and squadron marking, the fuel filler caps on the hull rear deck which were an identifying feature of the Sherman V and hull rear including the unit serial number and Arm of Service square.*

*Another view of Jordan's completed model. The thick mud was a mixture of plaster of Paris powder and static grass, darkened with earth-coloured pigments. It was worked into the tracks with an old paintbrush and then the excess was removed with a stiff brush. The final touch was an application of powdered graphite which was then buffed in select spots. The bodies of the crew figures are from Ultracast with heads produced by Hornet Models.*

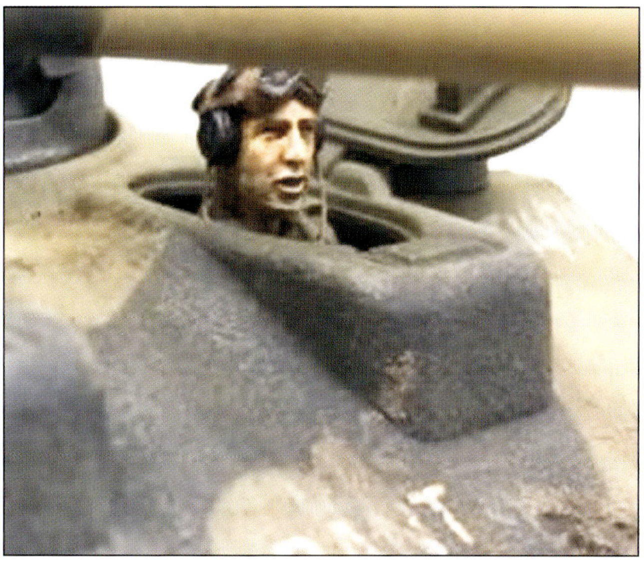

# SHERMAN V

## AND M-19 TANK TRANSPORTER

## 1ST CANADIAN ARMOURED BRIGADE

### 1/76 SCALE

NEIL CRAIG

*Neil's small scale masterpiece was built using the Matchbox M-19 Tank Transporter, long out of production, and the Sherman V resin kit from Milicast Models, both in 1/76 scale. Shown below is a photograph of the actual vehicle, a Sherman V named Cascade of The Ontario Regiment with one of the M-19 transporters of 1st Canadian Armoured Brigade's workshop.*

*Below: Matchbox's 1/76 scale M-19 Tank Transporter kit with the Milicast Sherman V (M4A4 Late Production) model at right. Cast in resin and made up of just four major parts, with a number of smaller details, this is still a reasonably detailed offering.*

The completed model on its scenic base. Both vehicles are finished in the Standard Camouflage Colour (SCC) 4 Stone and SCC 14 Black camouflage pattern that was applied to most of the brigade's vehicles before they left the UK. The figures are from Milicast except for the tank crewman in the black beret which is an AB Figures miniature. Note the field-modified fuel container storage on the tank's rear hull.

*Two views of the completed Sherman V. In addition to the stowage rack on the hull rear plate, Neil added small details such as the remains of the sunshield rail on the right side of the hull.*

Although the M4 Medium Tank, which the British and Commonwealth armies referred to as the Sherman, was one of the most important tanks of the latter half of the 20th century, with over 55,000 vehicles produced, it has only been little more than a decade since major model manufacturers began to take an interest in this versatile and sturdy armoured fighting vehicle.

Prior to that, the few kits that were on the market were either re-releases of very old models, which included inaccuracies in both detail and scale, or limited-production offerings in resin. The latter, although exquisitely produced, most often required a great deal of preparation and were usually characterised by a hefty price tag.

A number of small scale offerings were available from companies such as Fujimi, Hasegawa and the now defunct Italian company ESCI but these were essentially wargames models aimed at customers whose primary concern was not accuracy or detail but rather filling their armies with suitable tanks at a reasonable price in a short time. The venerable but extremely inaccurate Airfix Sherman fell into this category although, to be fair, it was first released in 1961.

The more recent model kits produced by companies like Tamiya, Dragon, Academy and Tasca are highly detailed replicas and increasingly feature photo-etched parts, detailed marking options and other aftermarket items as standard. Following the lead of the major manufacturers, specialised companies have developed additional items in resin, photo-etched metal and other materials and I have tried to include as many of these as space will allow. As with the previous books in this series I have chosen to concentrate on the most popular modelling scales of 1/35 and 1/48 although I have endeavoured to include models in other scales, both larger and smaller, where possible.

The reader should also be aware that for the most part I have included only versions which served with the Canadian, New Zealand and South African armoured units during the time period covered by this book. Due to reasons of space and the regular releases of new kits and accessories, this list is far from exhaustive and I would encourage readers to undertake their own research into the areas or variants that interest them most. An index of manufacturers can be found on page 64 of this book. Note that although some names, in particular Firefly, were probably not in use during the war I have employed them here as a matter of convenience.

## *TAMIYA INCORPORATED*

Beginning life in the immediate post-war period as a sawmill and timber supply company, which also produced wooden ship and aeroplane models as a sideline, Tamiya is today an international corporation. The company released a 1/21 scale Sherman M4 in 1963 and although a reasonable model for the time it would be regarded as a toy today. Nevertheless, by the early 1970s Tamiya was producing a large range of armoured vehicles, figures and accessories and was almost solely responsible for the rise in popularity of 1/35 scale. For their day the kits were highly detailed and accurate, offered extras such as crew figures, were relatively easy to assemble and, most importantly, affordable. The company's near monoply of the market remained unchallenged until the release of the first Dragon Models kits in 1987. In 2003 Tamiya began releasing a series of models in 1/48 scale and this line has been extremely successful for the company, combining the potential for a high level of detail without the expense of the bigger kits. The 1/48 scale offerings include an early production version M4, an M4A1 and a Sherman IC Firefly. It should be noted that only the latter has all the fittings necessary to build a specifically British tank. In 2014 Tamiya released a Firefly VC as a limited edition kit which was in fact the old Tasca model.

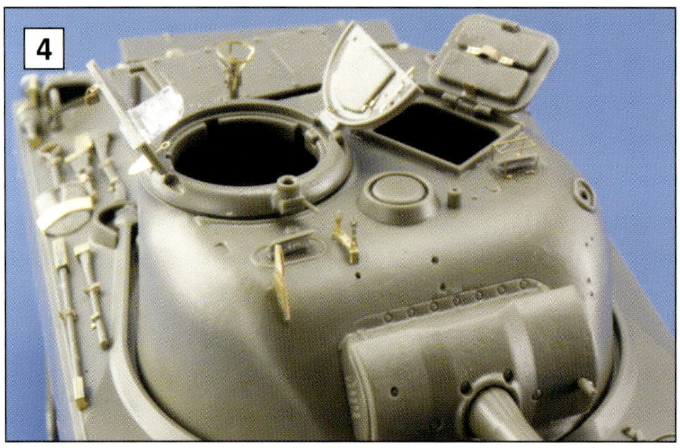


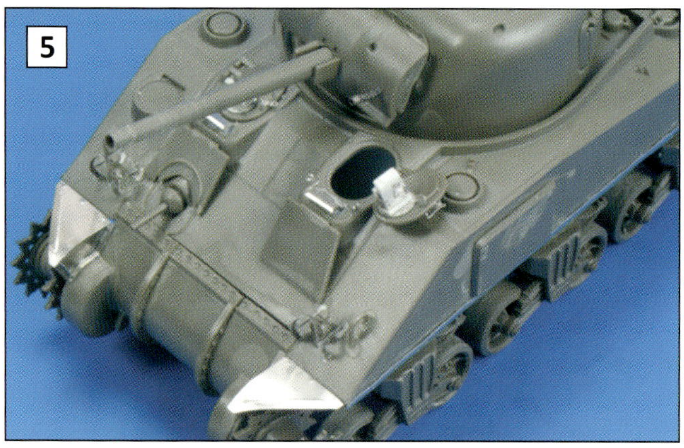


*Above: 1. Tamiya's 1/48 scale Sherman IC Firefly. 2. The M4A1 kit in 1/48 scale. 3. The 1/35 scale Sherman VC Firefly. All these tanks were employed by British and Commonwealth units in Italy. 4. Tamiya's 1/35 scale Firefly showing the turret and rear hull built with photo-etched and resin details from Eduard Model Accessories. 5. The 1/48 scale M4 kit completed with metal and resin parts from Royal Model. I should mention that this version was not allocated to Commonwealth units in Italy and I have included it to show the level of detail for which Tamiya is known..*

## DRAGON MODELS

Based in Hong Kong, this company offers sixteen construction kits in both 1/35 and 1/72 scale and twelve assembled and pre-painted 1/72 scale display models of the Sherman tank. Dragon Models was one of the first of the major manufacturers to market accurate, highly detailed replicas and their small scale releases are some of the best on the market. Dragon's 1/35 scale kits that represent Sherman tanks in British service in Italy include a Firefly VC, a Firefly VC with added .50 calibre machine gun and infantry figures and a Firefly IC Hybrid. There are also three kits of the Sherman III with cast and direct vision hoods, an early Sherman II, an M4A4 and an M4 with composite hull. The section on technical details and modifications which begins on page 49 will explain this further. The 1/72 range of construction kits contains a Sherman V with 60lb Tulip rockets, although these were not used in Italy, and a Sherman IC Firefly. In addition there are is an M4A1, an M4A4, an M4 with deep wading gear and a Sherman III. In the pre-painted and assembled armour series there are a number of models of the Firefly and Sherman III.

*Above: 1. Dragon Models' 1/35 scale Sherman III. 2. 1/72 scale Sherman V. The rockets were not used in Italy but are easily omitted from the build. 3. 1/72 scale Firefly VC kit. 4. 1/35 scale Firefly IC Hybrid with additions from Eduard Model Accessories. For some reason the armoured extension for the radio has been left off the turret bustle. 5. 1/35 scale Sherman VC Firefly kit built with upgrades from Royal Model. Below: Dragon Models' 1/35 scale Sherman IC with photo-etched brass detail.*

## ASUKA/TASCA MODELS

Rebranded as Asuka Model in 2013 this Japanese company has retained a number of the old Tasca Sherman models from its catalogue although the exceptional Sherman III kits with direct vision and cast hoods appear to have been dropped from production. In 2014 Tamiya released a re-boxed Tasca VC Firefly as a set with six infantry figures and there was some speculation at the time that they may have taken over all the Tasca moulds. However, the current Asuka catalogue shows the 1/35 scale Sherman V and the IC and VC Firefly as still available. The company's Sherman II depicts a tank with the early direct vision slots in service in North Africa but could, with very little modification, represent a tank in use in North-west Italy. Asuka also produces a number of Sherman suspension upgrades coupled with Modelkasten flexible tracks. These models are some of the most accurate Sherman replicas available.

*Below: Asuka's Sherman V kit assembled and painted as a British tank operating in North-west Europe. This variant, complete with turret-mounted smoke dischargers, was the most important tank in use with Canadian units during the Italian campaign. At right: The M4A1 kit built without any additions. Although marketed as a US Army tank these vehicles were employed by the British as the Sherman II.*

## RB MODEL

RB Model from Poland produce a range of milled aluminium and brass gun barrels and resin accessories for 1/35 scale armour models. Shown below, from left to right, are ammunition for the Firefly's 17pdr gun in 1/48 scale, highly detailed and accurate British wireless antenna fittings in 1/35 scale and the detail set for a .50 calibre machine gun with, inset, appropriate spent casings in 1/35 scale.

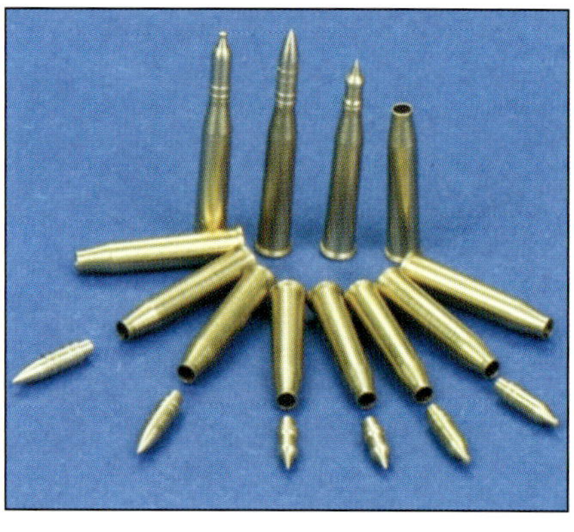

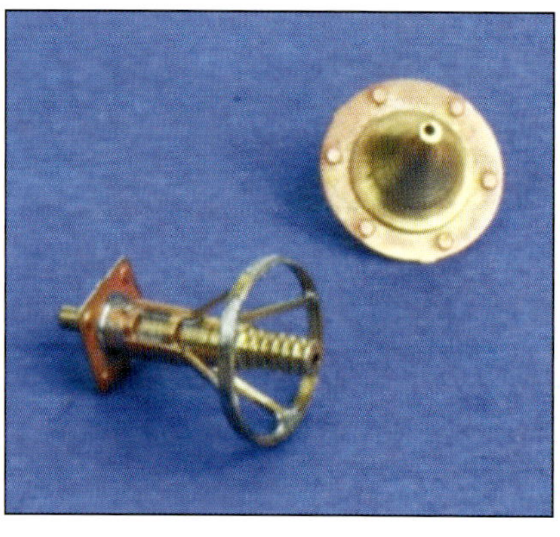

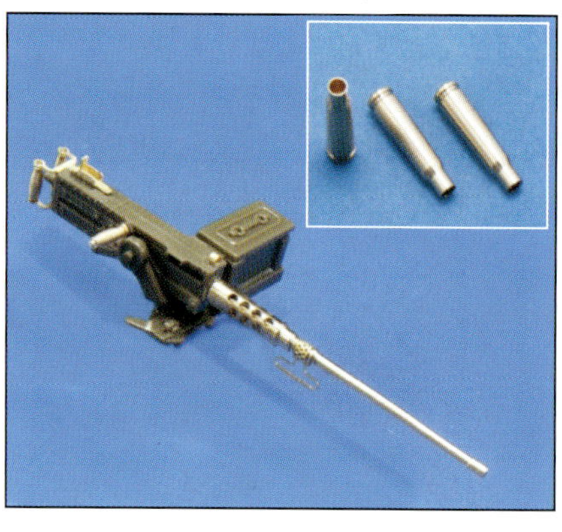

## EDUARD MODEL ACCESSORIES

Founded in 1989, this Czech company produces high quality photo-etched brass detail sets and a number of scale models. Currently available are photo-etched detail sets for Tamiya's VC Firefly in 1/35 scale and IC Firefly, M4A1 and M4 in 1/48 scale. In addition there are photo-etched sets for the Dragon IC Firefly and Sherman V in 1/35 scale although the Dragon Sherman III detail sets have unfortunately been discontinued at this time. Eduard also offers generic brass Sherman headlight guards in 1/35 scale and sets of vinyl masks to assist with painting the wheels of Tamiya and Dragon Sherman kits in both 1/35 and 1/48 scale. The company also produces a complete kit of the M4A1 with photo-etched details and extensive marking options, although these are all for tanks of the US Army.

1

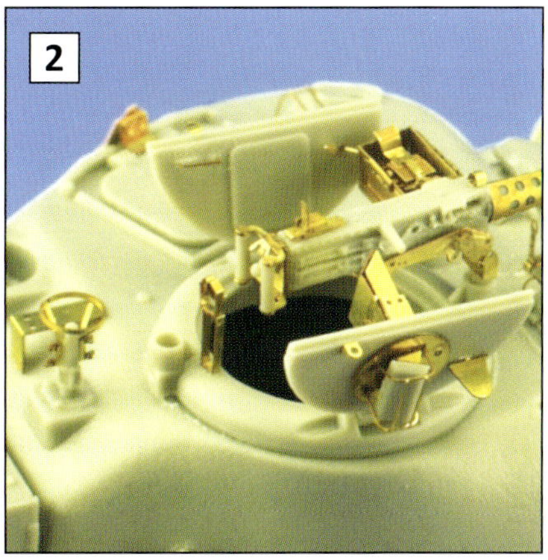
2

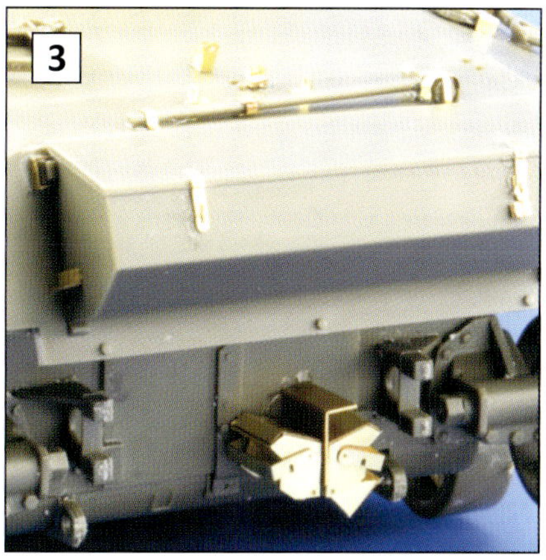
3

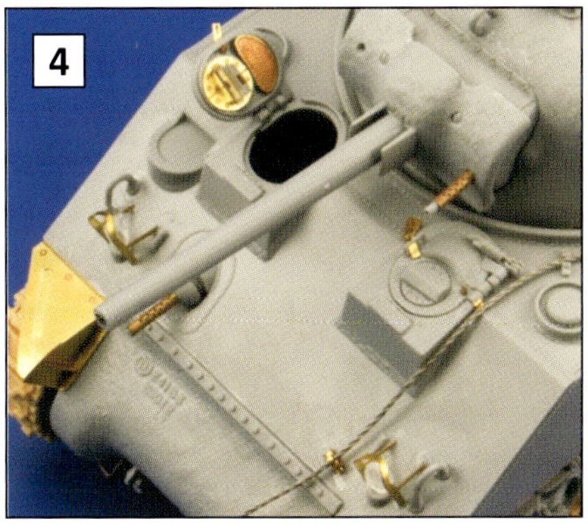
4

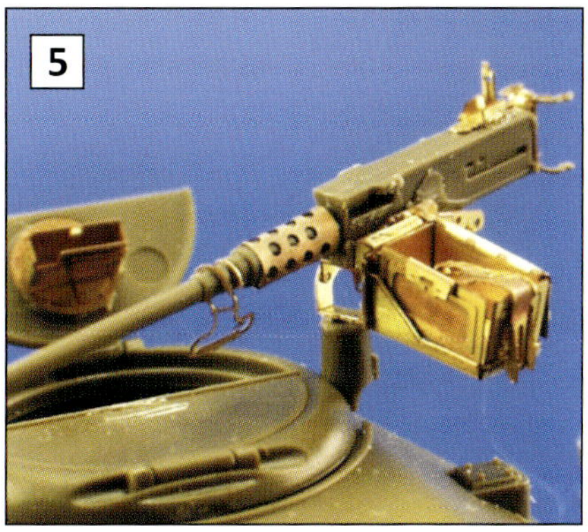
5

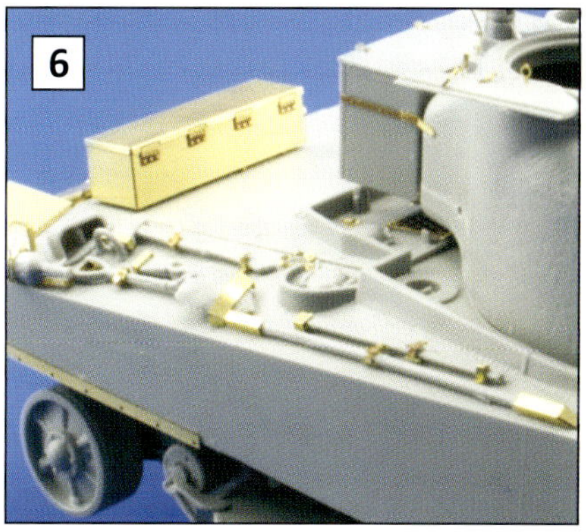
6

*Above, left to right: 1. Tamiya 1/48 scale Firefly VC built with the Eduard photo-etched set. 2. Dragon Models' 1/35 scale Firefly VC. 3. Tamiya 1/35 scale Firefly VC. The complicated-looking pieces of equipment fixed to the lower hull between the rear axles on the engine access doors are electrically-operated smoke emitters. The right-hand box is shown in the open position as it would appear after discharging its smoke candles. 4. Dragon Models' M4A2. Although the detail set for the Sherman III has been discontinued, this is a viable alternative. 5. The Tamiya 1/48 scale M4 Sherman. 6. Dragons' 1/35 scale Firefly IC kit.*

## HAULER

Based in the Czech Republic, this company produces accessory sets in photo-etched brass and resin in 1/48, 1/35 and even 1/87 scale. At the time of writing the company's catalogue offered photo-etched sets for the Sherman M4, M4A1, M4A4 and a specific British set. It should be noted that most of these sets are intended to upgrade US Army and Marine Corps models of the Sherman and additional items would be required to complete British versions. Hauler also produces an accessory set for the 1/72 scale Armourfast M4 model. It appears that the 1/48 scale Duckbill track extenders shown here are no longer in production although some retailers do have stock at the time of writing.

*Above, from left to right: Photo-etched end connectors fitted to T48 rubber chevron tracks. The Hauler detail set for the Tamiya M4. The photo-etched detail set for the Tamiya Firefly IC. All the items are for 1/48 scale models.*

## ABER

This Polish company has been manufacturing modelling upgrade sets since 1995, working in photo-etched brass, milled aluminium and brass, stainless steel and even wood. For some time now Tamiya has included a number of Aber products with their models. Although the company's list of accessories is extensive, those items dedicated to the Sherman are quite limited. Shown below are, from left to right, parts of the 1/48 scale upgrade set (1), the 17pdr barrel and muzzle brake in turned aluminium and brass (2) and the brass detail set for a .50 calibre machine gun both in 1/35 scale (3).

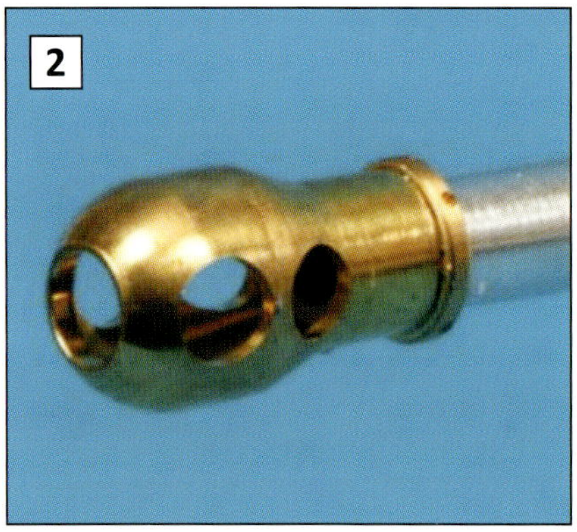


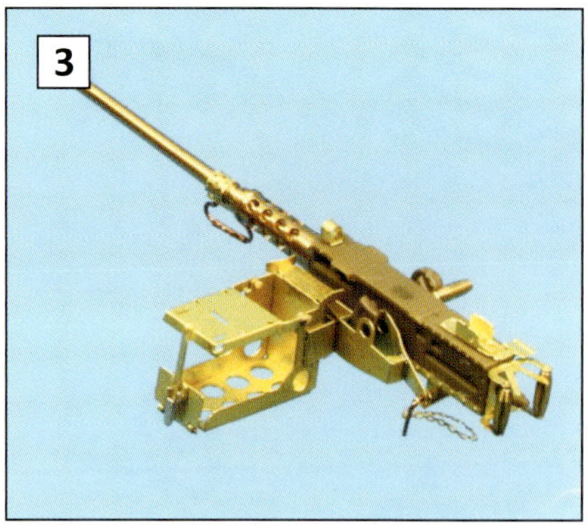


## E. T. MODEL

E.T. Model, a relatively new manufacturer from China, produce upgrade sets in 1/72 and 1/35 scale in brass and resin. The detail set marketed for the Dragon IC Firefly based on the Sherman I Hybrid hull shown here is the only product from the company's catalogue for a specifically British tank and contains parts made from resin, photo-etched brass and aluminium. Most of these parts could in fact be used for the Firefly IC based on the welded hull and Firefly VC and of course the Sherman I and Sherman V. The company also offers an upgrade set for the Dragon Models' US Army M4 and M4A1.

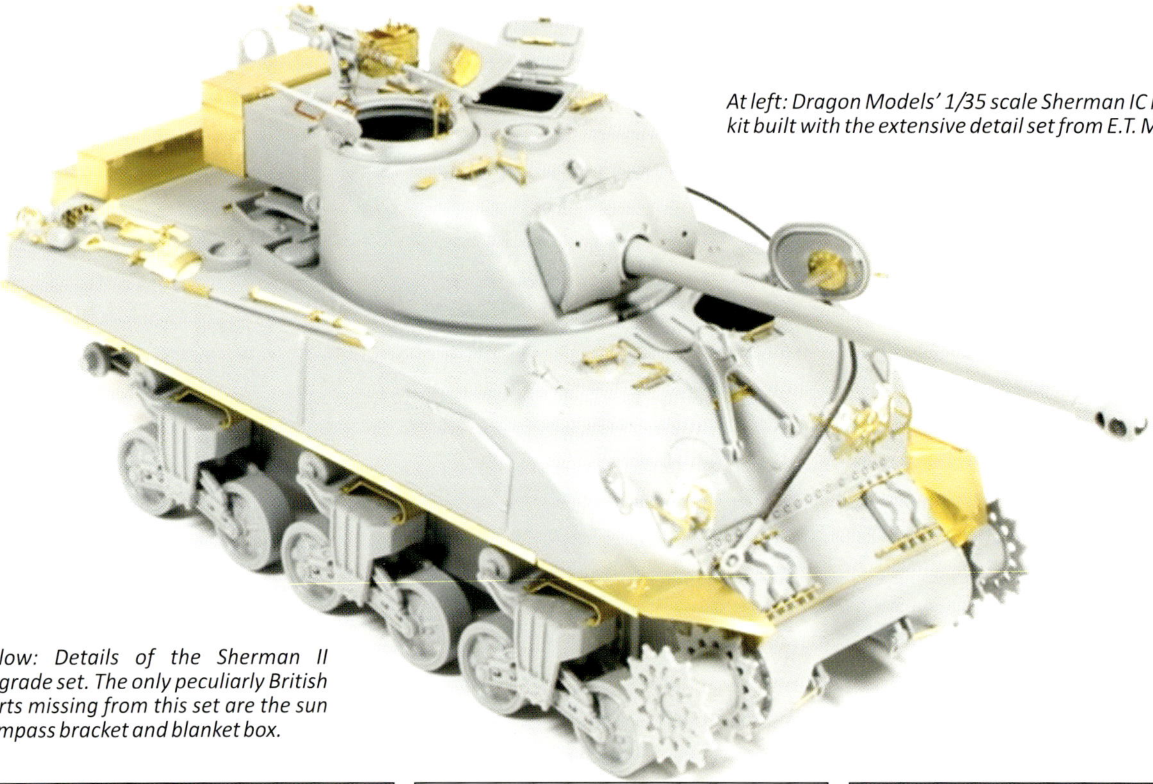

*At left: Dragon Models' 1/35 scale Sherman IC Hybrid Firefly kit built with the extensive detail set from E.T. Model.*

*Below: Details of the Sherman II upgrade set. The only peculiarly British parts missing from this set are the sun compass bracket and blanket box.*

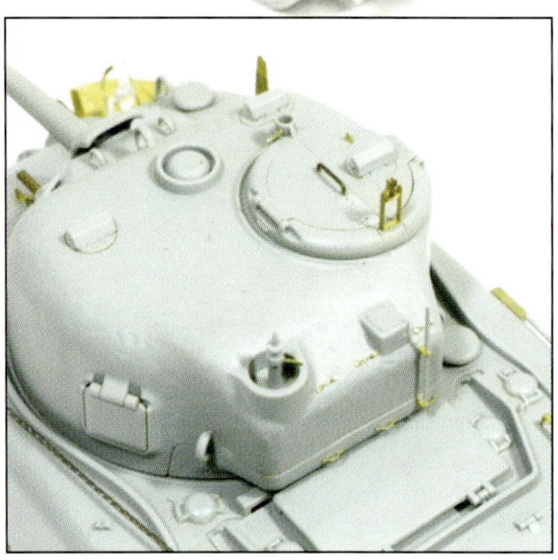

## ACCURATE ARMOUR

Based in Glasgow, this company produces a line of accessory sets and complete kits with a heavy emphasis on the British and Commonwealth armies, of both the 1939-45 period and modern day. Upgrade sets are available for the Firefly and Sherman I with spoked and pressed metal wheel sets, 17pdr shells and an assortment of ammunition containers. They are also a major outlet for the products from Canadian company Ultracast. Shown below are the detail set for the 1/35 scale Firefly, a Porpoise ammunition sled and a close-up of one of the wooden 17pdr ammunition crates.

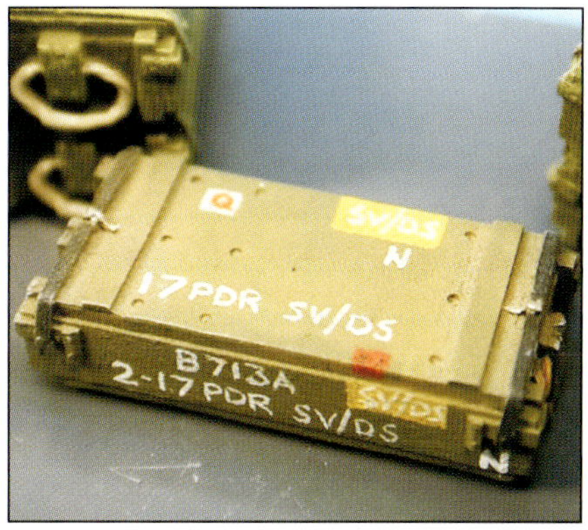


## ROYAL MODELS

This Italian company has been producing high quality accessories and aftermarket parts since the early 1990s under the guidance of its founder, Roberto Reale. The current catalogue includes complete upgrade sets in 1/35 scale for the Dragon Firefly and in 1/48 scale for the Tamiya M4. There are also sets for small details such as the bogies and hatch periscopes as shown here.

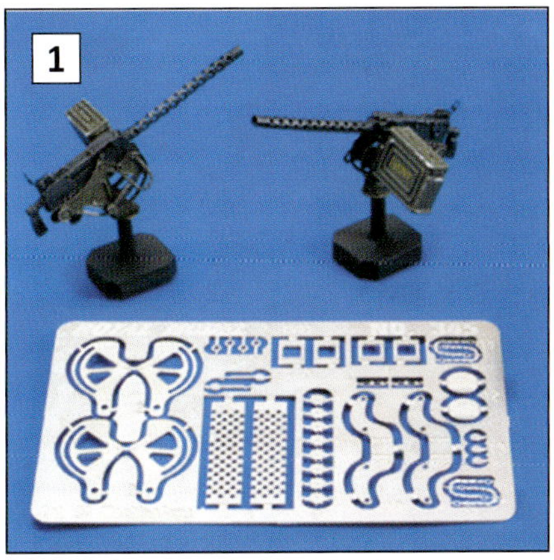


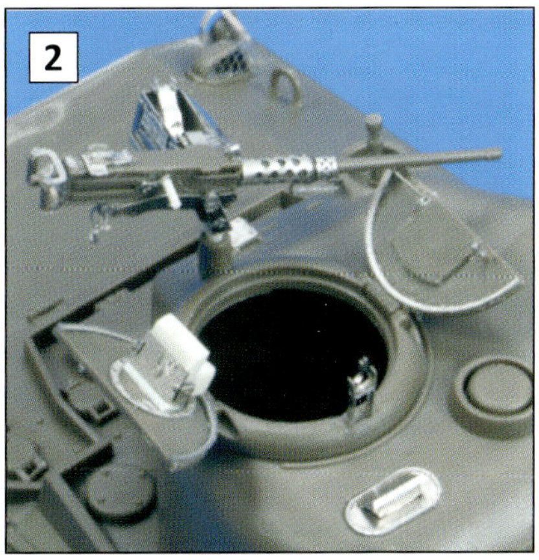

*Above: A selection of products from Royal Models. 1. The .30 calibre machine gun complete kit in 1/48 scale. 2. Part of the upgrade set for the Tamiya M4 model. All the parts shown here would be suitable for any 75mm-armed Sherman tank. 3. Detail of a built and painted 1/35 scale later model Sherman turret with loader's hatch showing Royal Models .50 calibre ammunition belts.*

## VOYAGER MODEL

Voyager have been manufacturing upgrade sets for scale models since 2003 with the release of their first photo-etched brass set for 1/35 and 1/48 scale armour. The highly detailed set for the 1/35 scale Firefly includes resin periscopes, metal towing cable and even a lucky horseshoe in brass. It should be mentioned that the fittings for the so-called Houseboat, a canvas cover that was intended to make the tank look like a lorry, were not seen on tanks operating in the Mediterranean. The detail set for the Dragon Sherman III does include items that were commonly seen on Shermans used in North Africa and Italy and many pieces would be suitable for other versions of the tank.

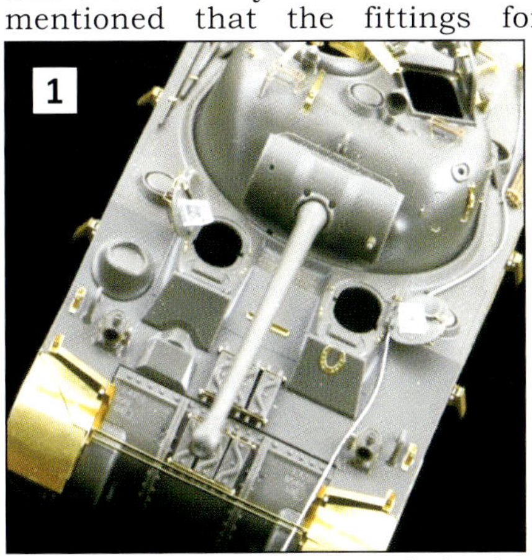


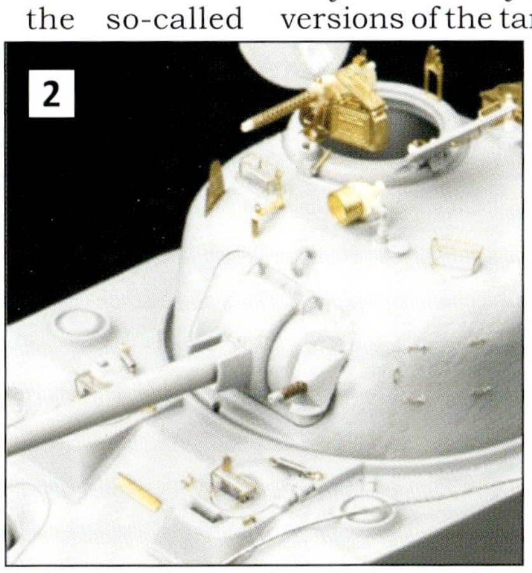

*Above, left to right: 1. Sherman Firefly detail set in 1/35 scale. The Houseboat fittings are the metal arms that extend over each front fender and the large brackets on the hull side. 2. Some of the photo-etched details applied to the Dragon Sherman III. 3. Photo-etched brass details for the Tamiya 1/48 scale M4 model.*

## *MARKING OPTIONS*

No model is complete with out a full set of markings and contrary to what many Axis modellers might suggest, British and Commonwealth tanks are much more than an endless procession of Olive Drab, or perhaps more correctly Standard Camouflage Colour SCC 15. Although most, if not all kits, come with a set of waterslide transfers, or decals to our American readers, many are limited and those produced by specialist manufacturers often allow for a more unusual, even exotic, version of a particular vehicle.

Star Decals, formerly Bison Decals, produces an extensive range of markings for Canadian, New Zealand and South African units and modeller Garry Boxall used them with his Sherman III build which is shown below and also in the Model Showcase section. I have collaborated with the firm's owner, Johan Lexell, on several projects and can recommend these sets for their accuracy. Archer Dry Transfers produce a large range of generic British markings but they were closed when this book was written.

*Above: A selection of Star Decals' waterslide transfers in 1/35 scale for South African (1), Canadian (2) and New Zealand (3) Shermans operated in Italy. Garry Boxall's beautifully finished Sherman III of 20th New Zealand Armoured Regiment which incorporates one of Star Decals' sets*

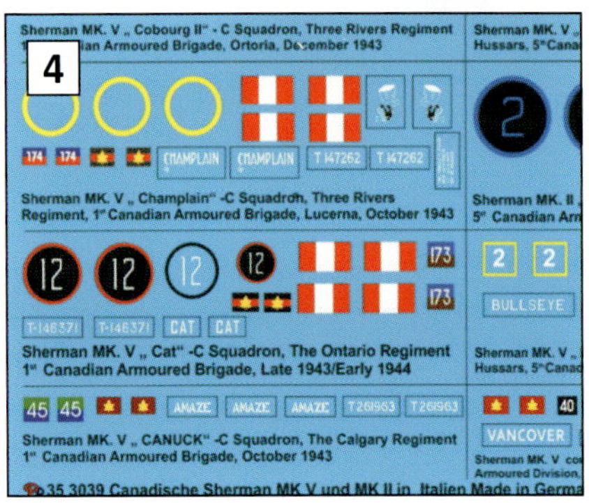


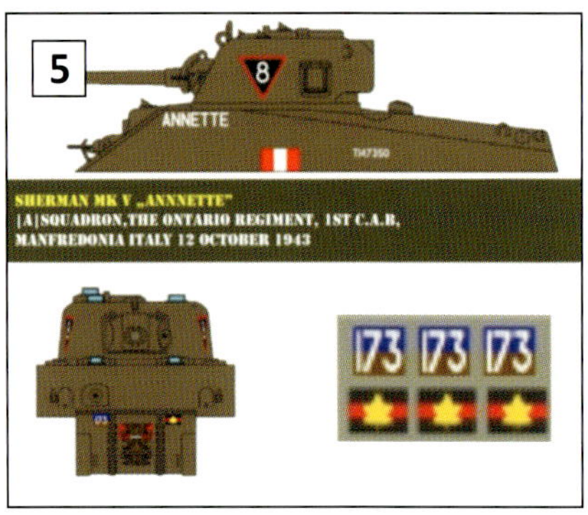


*At left: Part of the Canadian Sherman transfer set from Peddinghaus Decals (4). I was delighted to discover that this German company has been using my artwork as instructional material with some of their offerings. 5. RC-Tank, another German company, produces marking sets which include paint masks and waterslide transfers. The latter are used for the more detailed pieces such as the formation badge and unit serial numbers shown here.*

From early 1942, when the first production Sherman tank left the assembly line, until the end of the war in Europe in May 1945, the basic design underwent a number of major and perhaps hundreds of minor design modifications. Added to these were the many adjustments deemed to be conducive to the tank's operational efficiency and the comfort of the crews which were made to the vehicles shipped to Britain under the Lend-Lease progamme. These included a stowage bin or blanket box fitted to the turret rear, a bracket to hold the sun compass, the fitting of two 4in smoke dischargers on the turret side, rails to hold a sunshield on early deliveries and brackets to hold spare track links. Later additions were fire extinguishers clamped on either side of the hull towards the rear and, on some tanks, a smoke emitter fitted to the engine access doors. These and other modifications are discussed in the following text and shown in the illustrations in the Camouflage & Markings section.

Therefore, this section is not intended to be a definitive history of the development of the Sherman tank but will hopefully serve as a useful spotter's guide and also explain some of the terminology which many readers may find confusing. At the outset it should be mentioned that the practice of naming tanks was an entirely British convention and there is some evidence that Prime Minister Churchill suggested the policy of naming American tanks in British service after Civil War generals. The M4 medium tank was seldom, if ever, referred to as the Sherman by the US Army during the war. In any case the name Sherman has become so well known that I have used it here and throughout this book as a matter of convenience. Although the term Firefly, which is generally used today when referring to the 17pdr conversions based on the Sherman, was probably never employed during the war I have utilised it here to avoid the repetition of the long official designation. Similarly, terms such as large hatch or high bustle are almost certainly modern-day inventions but as they adequately describe particular parts or elements of these tanks I have continued to employ them. It should be noted that assembly modifications made to a particular model, for example the introduction of the M34A1 gun mantlet or the fitting of applique armour, were applied to most tanks that were in production at the time and sometimes retro-fitted to older vehicles. Listed here are the major models and variants of the Sherman tank employed by the Canadian, New Zealand and South African armies during the Italian campaign. The tanks are referred to here using their British designations followed in parentheses by the US Army classification.

**Sherman I (M4 Medium Tank).** Although the detailed tank strength lists complied by 15th Army Group, which controlled all the armoured units serving in Italy, indicate that no 75mm-armed Sherman I tanks were on hand I have included the type here as many of the design features, such as the turret and suspension, were incorporated into the other versions. Similarly, the Sherman I and Sherman I Hybrid (M4 Composite Hull) provided the basis for the 105mm-equipped Sherman IB and the Sherman IC Firefly described below.

Produced from July 1942 until January 1944, Britain received a total of 2,096 Sherman I models, including the composite hull versions, and all were powered by the Continental R-975 air-cooled radial engine. The initial welded hull (A) was designed with the front glacis plate inclined at an angle of 57 degrees and featured two box-like extensions, referred to as hoods, which incorporated hatches for the driver and lap gunner (1). The first production versions were fitted with direct vision slots (B) in front of the two hull positions but these were later welded welded shut and eventually removed from production and replaced with fixed periscopes (C). An antenna bracket was welded onto the right hand side of the glacis (D) and the shape and construction of these varied between manufacturers. On early models the final drives and differential were protected by a three-piece armoured cover that was bolted together (E) but this was later replaced by a cast, single-piece variant. The latter was produced in two versions with the earlier type being more rounded (F). On the engine deck a large air inlet cover was situated just behind the turret with fuel filler caps located on either side. Towards the rear of the hull a door which opened forward allowed access to the engine. The hull rear

## Notes

1. Or assistant driver in US Army terminology.

plate could be distinguished by an inverted u-shaped cut-out which partially covered the air filtering system (A) and this was an identifying feature of the type. Two hinged doors at the rear of the hull gave access to the engine. Towards the end of 1943, 1in thick applique armour plates were welded to the hull sides (B) and examples of these are shown in the accompanying photographs and also in the Camouflage & Markings section. The large-hatch version of the M4 tank, with the hull glacis angled at 47 degrees, was used for the Sherman IB 105mm-armed vehicle described below.

The chassis incorporated the Vertical Volute Suspension System (VVSS) which was made up of a drive sprocket (C), a rear idler and three two-wheel bogies on either side (D). The latter had been developed from the versions used with the M3 and the variant with offset roller and skid are the most common type encountered in Italy (1).

The turret retained its basic shape throughout production of the 75mm-armed versions and featured a cast cupola with a split hatch for the commander (E), a centrally-placed ventilator (F), a loader's periscope and a pistol port on the left-hand side (G). In April 1943 it was decided to delete the pistol port from production but this was so unpopular with the troops that it was reinstated in the following July, although not before a significant number were welded closed. The lifting rings welded to the turret sides towards the rear were initially located near the top edge but by the end of 1942 they had been moved closer to the bottom. A third lifting ring was located at the front of the turret above the gun mantlet. To accommodate the turret traverse mechanism it had been necessary to cast the turret with thinner armour on the front right side and to protect this vulnerable area applique armour was incorporated into production in spring 1943. Modification kits were provided to depots and units in the field but within months thickened armour had been incorporated into the casting process, producing a turret with a pronounced bulge on the right side which ran from the gun mantlet along the turret side to about halfway towards the rear.

The overhang at the rear of the turret is commonly referred to today as the bustle although it should be stressed that the term is a modern invention and was certainly not used during the war. The first production models were produced with a so-called low bustle but from late 1943, in an effort to improve clearance for the larger hatches, which were then being introduced, the overhang was raised slightly resulting in the high-bustle configuration.

Early production vehicles were fitted with the M34 gun mantlet (H) until its replacement with the M34A1 version (I) in March 1943. From late 1942 the early gun mantlet was produced with a protective armoured half-sleeve and at about the same time a cast armoured shield was introduced which covered the coaxial machine gun. The latter was also

## Notes

1. The later Horizontal Volute Suspension System (HVSS) need not concern us here.

manufactured as a modification kit and first shipped to units operating in the Mediterranean in the spring of 1943.

In British vehicles the radio was moved from the hull to the turret and this arrangement later became standard. The No. 19 wireless set consisted of two radios and each required its own aerial. The A aerial was installed in a socket on the left rear of the turret while the B aerial was also mounted in a smaller socket just right of centre (A).

From December 1943 turrets featured an oval-shaped hatch for the loader and a cupola with vision blocks for the commander but I have been unable to find a single image depicting these modifications on a 75mm-armed Sherman operating in Italy. The composite hull, with its cast front (B), was seen as a solution to production delays and the high costs involved with the all-welded process. As the welding of the hull front accounted for most of the production time, the firm of Chrysler suggested that a cast front be fitted to the M4 hull and in August 1943 full-scale production began. Many, if not all, these tanks featured improvements which were being incorporated into production at the time including applique armour on the hull sides (C), a gun travel lock on the glacis and the cast differential housing. Given the production date, a number would have been fitted with the turret without the pistol port. Confusingly, the US Army did not differentiate between the Sherman I and Sherman I Hybrid and the term composite hull, although descriptively accurate, was not an official designation. The first fifty production vehicles were built with the small hatches of the M4 but the few Sherman IC Firefly conversions, which began to equip Commonwealth units in Italy in mid-1944, seem to have all been based on the large hatch hull versions.

*........................text continued on page 53*

*Above: Photographed in early 1945 during the fighting along the Senio River, possibly near Sant'Alberto, this Sherman I Hybrid was operated by the North Irish Horse, a British regiment. There is no record of M4 Shermans in use by 15th Army Group units in the strength returns that are available, other than 17pdr conversions, and they may have arrived after April 1945. Points to note are the large hatches with their periscope guards, the turret stowage bin and the gun travel lock. The small stub welded onto the hull below the latter is an identifying feature of composite hull vehicles although I can offer no explanation as to its purpose. Note also the command antenna on the hull front next to the lap gunner's position.*

*A line up of 105mm-armed Sherman IB tanks of 11th South African Armoured Brigade photographed in April 1945. Note the opened loader's hatch, commander's vision cupola and the distinctive gun mantlet of this type. These tanks, and the brigade's Fireflies, were removed from the squadrons and concentrated in the headquarters companies of the brigade's regiments, which explains the different markings seen here. Below: Details of the Sherman IB. A. The commander's cupola and the loader's hatch. B. The 47-degree glacis plate, the so-called large hatches and sharply-contoured cast transmission cover. C. Field modifications made by 4th New Zealand Brigade including a fuel can rack and small handles welded to the hull side. D. The gun mantlet with its canvas cover and the large hatches in the open position.*

A

B

C

D

.........................*text continued from page 51*

**Sherman IB (M4 with 105mm howitzer).** Designed as an infantry support and assault vehicle, this model was armed with a modified version of the US 105mmm howitzer. Unlike the initial M4 design, the hull featured the so-called large hatches with the glacis angled at 47 degrees and, in profile, was very similar in appearance to the later M4A3 model. The engine deck and rear hull, including the exhaust system and engine access doors, remained unchanged.

A prominent design feature was the high bustle turret which incorporated an escape hatch for the loader and an additional ventilator fitted toward the rear of the turret roof.

Early production versions were fitted with the standard commander's cupola but later examples had the all-round vision model and it seems that all the tanks that served with British and Commonwealth units in Italy were equipped with the latter. Second generation production versions featured an anti-aircraft machine gun pintle mount and stowage fittings welded to the rear of the turret.

The British, principally Montgomery, were reluctant to take this version on but were compelled to accept it due to a shortage of 75mm-armed tanks and a total of 593 vehicles were eventually received. Beginning in October 1944, Sherman IB tanks were allocated to British, Canadian, New Zealand and South African units operating in Italy.

**Sherman IC Hybrid (M4 Composite Hull).** Initially rejected as impractical and only grudgingly put into production as a stop gap measure, the Sherman tanks fitted with the British 17pdr gun were almost certainly the most successful Allied armoured vehicles produced during the war (1).

Conversions to mount the larger gun in place of the 75mm weapon were carried out using the Sherman I, Sherman I Hybrid and Sherman V (M4A4) models and they were identified by the C suffix. The suggestion that 17pdr guns were fitted to other versions, while persistent, is incorrect.

In order to mount the gun in the Sherman turret it was turned on its side and adapted for left-hand loading. As the breech of the main gun now filled much of the turret, and allowance had to be made for the gun's recoil, an armoured extension was fixed to the turret rear to hold the wireless that was normally positioned behind the commander and loader. This armoured box, which also acted as a counterweight, is a distinguishing feature of these vehicles. As the loader's access to the turret hatch was now obstructed by the gun breech it was necessary to cut a second hatch in the turret roof. This should not be confused with the loader's hatch incorporated into M4 production from December 1943 as mentioned earlier. To increase the amount of ammunition carried, both the lap gunner's position and the hull machine gun were removed and the aperture covered with a thick metal plate. The gun mantlet was completely redesigned.

The turret was mounted on the Sherman I Hybrid hull as described above but many of these tanks were withdrawn from service on the Italian Front when increasing numbers of the Sherman VC (see below) became available.

## Notes

1. The abbreviation 'pdr', for pounder, as applied to British and Commonwealth guns, refers to the weight, in pounds, of the projectile.

*At left: The modifications made to the Sherman turret to mount the 17pdr gun including the installation of a hatch for the loader and the turret rear extension. The vehicle's radio can be seen inside the latter which is yet to have its top bolted in place. As shown by our photograph the conversion process involved a considerable amount of cutting and welding and it is likely that these tanks were repainted in Standard Camouflage Colour (SCC) 15 Olive Drab. Note that the pistol port has been welded shut.*

*At right: 6th South African Armoured Division apparently received a number of 17pdr-armed tanks in late 1944 on a trial basis and this Sherman IC is probably one of those. Although it is difficult to tell from this image the barrel of the main gun appears to have been painted in some kind of disruptive pattern which was to become common later in the campaign. The A Squadron tactical sign may be in white denoting the division's unbrigaded reconnaissance regiment, The Natal Mounted Rifles, retained its tanks until late 1944 when it was effectively converted to an infantry unit. A number of the Sherman IC vehicles were still with the division when the war ended and took part in the victory parade at Monza.*

*At left: Sherman II tanks of B Squadron, 3rd Armoured Reconnaissance Regiment (The Governor General's Horse) photographed in July 1943 when the units of 5th Canadian Armoured Division were inspected by HM King George VI. The curved hull front with its distinctive ventilator cap is clearly visible here. Note the small bracket on the turret side which can also be seen in the photograph below. These tanks were almost certainly painted in US Army Olive Drab and one account states that they were cleaned with petrol for the King's visit which would have definitely given them a very dark appearance. The tank's name, Bull, is probably rendered in white, the colour used to denote an unbrigaded regiment.*

*At right: Another Sherman II of The Governor General's Horse photographed during the summer of 1944. After the British, 5th Canadian Division was the most important user of this version having forty-six examples on hand at one time. The vehicle shown here is a later production model with turret and hull applique armour, a gun travel lock and cast transmission cover. Note the small brackets on the turret and hull sides, presumably to hold foliage camouflage. In the original image the name Chimpanzee can be discerned on the hull side, painted on the foremost applique armour plate.*

**Sherman II (M4A1).** This version was the first model of the Sherman to enter British service and is easily distinguished by its curved, cast hull. Although the hoods for the driver and lap gunner were contoured into the hull, the outward-opening hatches were identical to those of the Sherman I.

Initial production versions featured the same direct vision slots that were fitted to the M4 and although these were officially dropped from production as early as June 1942 it was many months before the redesigned hoods made an appearance. The engine access doors and air filters were also the same as those used with the M4 and the rear hull plate featured a very shallow cut out.

Applique armour, when fitted to the hull, was either made from rolled plate or applied in sections to conform to the contour of the body. From October 1943 hulls were cast with thickened areas in place of the applique plates.

From January 1944 a small number of M4A1 tanks were built with the so-called large hatch and some of these were delivered to Britain where they were converted to Duplex Drive tanks. All were equipped with a Continental R-975 air-cooled radial engine.

5th Canadian Armoured Division had forty-six of these tanks on hand in September 1944 and they were still with the division in February 1945 when the Canadians left for North-west Europe. Examples are shown in the Camouflage & Markings section of this book. The Sherman II hulls that were used by 28 Assault Squadron New Zealand Engineers as fascine carriers may have come from British 7th Armoured Brigade but at least one was based on the hull of a Sherman IIA.

**Sherman IIA (M4A1 76).** In an effort to equip their armoured units with a more powerful gun the US Army proposed that the turret from the abandoned T23 medium tank program, complete with its 76mm gun, be fitted to an existing hull and they eventually settled on the M4A1. The Pressed Steel Car Company was the sole manufacturer of these tanks as this was in fact the only firm that would undertake the project and guarantee delivery of the first 160 examples by January 1944.

The T23 turret, while compatible with the M4 hull, was a complete redesign although it did retain some familiar features such as the periscopes and pistol port. It was fitted with the same vision cupola used with the Sherman IB as mentioned above while the loader's hatch was the standard commander's cupola that was installed in the M4 turret.

A large number, perhaps more than 400, were produced without the prominent ventilator on the turret rear. In an effort to standardise production the split-hatch cupola was replaced in August 1944 by an oval hatch similar to that introduced into M4 production from December 1943

........................text continued on page 56

*This Sherman IIA of The Queen's Bays, a British regiment, shows several important features of the type including the large hull hatches, the commander's vision cupola and the split loader's hatch which was fitted to early production vehicles. Note the prominent ventilator cover on the turret rear and the remains of the sandshields with which most of these tanks were initially fitted.*

*6th South African Armoured Division was the largest user of the Sherman IIA during the Italian campaign and its regiments began replacing some of their 75mm-armed tanks with the type as early as November 1944. Photographed in April of the following year, the tanks shown here, from the division's Special Service Battalion, all have the earlier split loader's hatches while the hatches of the commander's cupolas have been modified to open towards the rear. Note that the tank in the foreground has the early production turret without the ventilator cover. A single Firefly can see seen in the background on the right side.*

.........................text continued from page 54

## Notes

1. Indeed, with the exception of the almost 500 examples which were employed by the US Marine Corps, all the production M4A2 models were exported under the Lend Lease programme.

although it was slightly bigger. The new turrets had a periscope for the loader and a machine-gun pintle was added between the commander and loader's positions. It has been suggested that many of the 76mm-armed tanks allocated to Britain between August and October 1944, most of which went to 15th Army Group in Italy, were fitted with the newer turret but most available photographs show the earlier version. In addition it appears that the South Africans, who were the largest users of the Sherman IIA in the Mediterranean, reconfigured the commander's cupola to allow the hatch to open towards the rear. Examples of these vehicles can be seen on pages 25 and 26.

The hull was based on the so-called large hatch version of the M4A1 with redesigned front contours and slightly angled rear plate. Internally the hull featured wet stowage ammunition bins, a recent innovation, which required that the tank's ammunition supply be relocated from the sponsons to the hull floor. The principle of wet stowage worked by having the ammunition racks surrounded by a jacket containing a mixture of Ethylene Glycol and water. In the event of a shell fragment penetrating the jacket the liquid would, in theory at least, extinguish any flame before it could ignite the ammunition. But the almost universal practice of having extra ammunition lying about the turret effectively rendered the whole system useless. As mentioned earlier, 6th South African Armoured Division employed a large number of these tanks, reporting on 30 December 1944 that a total of 134 were on hand. However, they seem to have suffered an inordinate number of casualties during the battles of spring 1945 and just six were operational when the war ended in May.

**Sherman III (M4A2).** The welded hull of this model was similar in appearance to the early Sherman I with the most obvious differences being the engine access doors on the rear deck behind the turret, the exhaust systems for the diesel engine and the rear hull plate which was much larger, without any cut out and was sloped towards the rear. When the direct vision blocks were dropped from production tanks were manufactured with angular, welded driver's and lap gunner's hoods and also cast hoods that were angled at the sides at the front. Both were identifying features of this model. All Sherman III tanks were fitted with twin General Motors diesel engines and indeed were often referred to in early British reports as simply 'Sherman, diesel'. This aspect contributed in no small measure to the tank's popularity with British tank crews who believed that they were less likely to burn when hit by an armour-piercing round. It should be mentioned that all the M4A2 tanks shipped to Britain were the so-called small hatch versions manufactured up to May 1944 and in numbers received they were second only to the Sherman V with a total of 5,033 being delivered (1).

In Italy 4th New Zealand Armoured Brigade was equipped exclusively with the Sherman III until the arrival of the 17pdr Fireflies and the Sherman IB.

.........................text continued on page 59

*One of the first Sherman III tanks delivered to 4th New Zealand Armoured Brigade while the regiments were training in Egypt, this vehicle is painted in a base coat of Light Stone. Note the early gun mantlet, cast hoods, three-piece transmission cover and sandshields, Note also the rails for the sunshield on the hull side and the stowage basket welded to the right side fender.*

*Above: Identifying features of the M4A2, or Sherman III. A. The first production driver's and lap gunner's hood with the direct vision block. B. The welded hood showing the hatch interior with periscope. C. The cast hood with its angular sides. D. The engine deck with access doors and filler caps. E. The distinctive rear hull plate. F. Many Sherman III tanks of 4th New Zealand Armoured Brigade were fitted with lengths of Bren Carrier tracks to protect the engine exhausts. Although it is difficult to be certain, this modification may have been unique to the Sherman tanks of the brigade's 20th Armoured Regiment. Below: A Sherman III of 4th New Zealand Armoured Brigade which displays almost all the early design features of the Sherman III tank photographed in Egypt and shown on the previous page. Note the pattern of disruptive camouflage painted in accordance with the April 1943 instructions.*

*Above: A Sherman V of 4/22 Field Regiment, the second senior artillery regiment of 6th South African Armoured Division, photographed near Imprunetta in August 1944. The protruding driver's and lap gunner's hoods with their distinctive under cut, the cast antenna bracket and bolted three-piece transmission cover are all identifying features of this model and are clearly visible here. Note also the additional radio antenna on the hull front and the Swabey sight fitted to the turret in front of the commander. Below: A. A closer view of the hood clearly showing the bulge of its forward edge. B. The raised radiator filler on the rear deck just below the blanket box.*

*Sherman VC tanks of 6th South African Armoured Division photographed in April 1945. The loader's hatch and rear turret extension, both identifying features of the 17pdr conversions, are clearly visible here. Note the fire extinguisher on the hull side and specially designed stowage box on the hull rear. These items are noticeably absent from the earliest arrivals. The vehicle in the foreground is also shown and discussed on page 25 of the Camouflage & Markings section.*

........................*text continued from page 56*

As mentioned above, tanks were modified to bring them up to British standards and a check-list dated 22 October 1943 shows thirty-five internal and external modifications that were to be made to M4A2 tanks by either the manufacturer or the Ford Motor Company's Chester depot before they were shipped to Britain. These included the welding of additional armour plates in front of the driver's and lap gunner's hoods, the fitting of sandshields, the addition of a towing shackle and pin, the installation of gas vent tubes, the fitting of guards for the periscopes, the addition of a mark traversing and elevating wheel, the modification of the recoil guard, gyro stabiliser and battery switch and the installation of a turret ventilation fan. This does not include the modifications that would be made when the tank arrived in the UK or Egypt.

**Sherman V (M4A4).** Although outwardly similar to the Sherman I, the M4A4 hull was built some 6in longer to accommodate the massive Chrysler 5-line engine. A slightly modified suspension was also required and the number of track shoes was increased from seventy-nine to eighty-three. As with the other models, this tank featured the early M34 gun mount until it was replaced in March 1943 by the M34A1, the version which is most often seen in photographs of the Sherman V. The early type, three-piece transmission cover was retained on all vehicles and identifying features were the protruding hoods which featured a slight undercut, a cast antenna bracket, the prominent radiator filler on the the rear deck below the turret overhang and the sloped rear plate. This model was by far the most numerous Sherman type in service with the British Army with 7,155 vehicles being supplied by the war's end. The Canadian 1st Armoured Brigade was equipped exclusively with Sherman V tanks and was in fact the first unit to employ the type operationally in July 1943 when C Squadron of the Three Rivers Regiment took part in the assault landing at Pachino Bay in Sicily.

**Sherman VC.** As mentioned above conversions to mount the 17pdr gun in place of the Sherman tank's standard 75mm weapon were carried out using the Sherman I, Sherman I Hybrid and Sherman V. Although these tanks were undoubtedly the most potent armoured fighting vehicle fielded by Britain during the war they were always in short supply and in addition the units operating in North-west Europe were given priority.

In Italy they were not in use until late 1944 and the 15th Army Group situation report for September of that year states that just twenty were being readied for issue to units while a further seventy-nine were in transit from Britain. These tanks were in fact Sherman IC models and the first Sherman VC Fireflies were not on hand until early in 1945 when they were for the most part concentrated in 4th New Zealand Armoured Brigade. Photographs show that a number were on hand with 6th South African Armoured Division but must have been received after April 1945 (1).

## Notes

1. At that time fifteen Sherman VC were in transit to Italy from Britain. There are accounts of the South Africans receiving a number of Fireflies in late 1944 on a trial basis but these are not shown in any RAC strength returns.

One further technical aspect that should be considered is the use of camouflage colours. By the time of the Sicily landings clear orders governing the application of camouflage were in place for units operating in the Middle East and those stationed in the UK. The question of just which colours and patterns were used in the subsequent fighting in Italy is complicated by the fact that the formations which served there were raised and equipped in both Egypt and Britain and to gain a clear understanding it may be best to start in October 1942 when the first Sherman tanks were allocated to British units in North Africa.

At about the same time, the Camouflage Directorate of Middle East Command completed a study on the effectiveness of the various colours and designs that had been used for vehicle painting up to that stage. The study acknowledged the limits of camouflage painting but declared that since vehicles must be painted it should be done in a manner that would make them as inconspicuous as possible.

In a complete reversal of existing policy the Directorate's study argued that not only should all vehicles of the same type or class carry an identical design, but that this should be consistent throughout the army. The basic colour should be one that could be used in the diverse types of terrain encountered across the Middle East, but most importantly in the desert areas where much of the fighting was expected to take place. The shade chosen for the base colour was a warm sandy pink, later designated Desert Pink No.Z.I., which was to be produced locally. It was considered that this colour best suited the desert background which varies depending on the time of day and atmospheric conditions. Over the base colour a disruptive pattern was to be applied and the preferred colour was Dark Green but Black, Dark Brown or Dark Slate were given as suitable alternatives (1).

The findings of the study were accepted and disseminated as a draft General Order (GO) in early October 1942. On 19 October an amendment was issued which stressed that any disruptive pattern was to be gently oblique to conform to the horizontal lines of the desert and that vertical lines were definitely not to be used. The new order also went on to explain that more of the lighter base colour should be visible and that as disruptive patterns with soft edges would defeat the purpose of the camouflage, hard edges were preferred. This instruction also contained a reminder that undercut portions of the vehicle were to be painted white to negate the effect of harsh shadows. One cannot escape the conclusion that the need to stress these points, all included in the original order, suggests that they were not being adhered to (2).

The October orders were ratified by GO 1650 of 4 December 1942 which included diagrams showing the authorised disruptive pattern from all sides, including a top view, and it was again stressed that these were to be strictly followed. All new and repaired vehicles were to be re-painted by Base Ordnance Depots before issue but commands and formations were to arrange for the painting of vehicles held by them. This resulted in the old base colour of Light Stone, a very light sand-yellow shade, remaining in use for some time and it is

## Notes

1. The disruptive colours were all taken from British Standard (BS) 381c of 1930.
2. Significantly, this new order called for the diagrams included with the original order to be returned for destruction. It would seem safe to assume that some of these patterns differed significantly from those issued with the 19 October 1942 amendment.

*Below: Photographed in October 1943 as it crossed the Volturno River near Grazzinise north of Naples, this Sherman III of 4th County of London Yeomanry, a British regiment, is painted in a textbook application of the April 1943 Light Mud and Black camouflage pattern. A detail taken from the template provided with the XIII Corps order, which is our reference for this scheme, is shown at left. Note the rail on the hull side for the sunshield. The name was deliberately deceptive and the sunshield was actually a canvas cover on a metal frame that was intended to disguise the tank as a lorry. The presence of the rail on Commonwealth tanks is a reasonable indicator of it being handed over from a British regiment.*

*At left: Captain W.H. Willis, a company commander of the 11th Armoured Regiment (The Ontario Regiment), photographed in February 1944 with one of the regiment's Sherman V tanks. The two-colour camouflage made up of SCC 4 and SCC 14, as described in the text, is clearly visible here with the dark disruptive running along the lower edge of the hull and turret. This undulating pattern is almost certainly what was referred to in the regimental diary as a 'broken line'. Note the very low contrast between the tank's base colour and the red of the RAC flash, visible just behind Captain Willis' left arm.*

*At right: A Sherman V of A Squadron, 14th Armoured Regiment (The Calgary Regiment) named Arrant, said to be the first Allied tank to land in Italy. The disruptive camouflage is far more pronounced here than in the example shown above, as is the contrast between the two colours, and other images show that the disruptive was extended to the turret roof. This tank was photographed several times and also filmed and from some angles what appears to be a thick, slightly lighter-coloured outline can be seen on the darker colour applied to the hull. Other tanks of this regiment had similar disruptive patterns with a distinct light outline which had a very soft edge.*

obvious that on a number of vehicles the suspension and wheels retained their coat of US Army Olive Drab, probably in an effort to conserve paint. In addition some tanks lacked the disruptive pattern. But by December 1942 photographs show a much more uniform appearance with the disruptive camouflage closely following the diagrams distributed with the October orders.

In the UK at this time the officially authorised camouflage scheme consisted of a base coat of Standard Camouflage Colour (SCC) 2 Brown with a disruptive pattern of SCC 1A Dark Brown although SCC 14 Black was an acceptable alternative (1).

In early 1943 units of the Eighth Army crossed into Tunisia and it was immediately apparent that the desert camouflage was not only unsuitable but completely counterproductive against a backdrop of shrub-covered hills and olive groves.

In April 1943 a completely new set of instructions were issued and a new base colour referred to as Light Mud (2), was authorized and any disruptive pattern was now to be painted using Black, the patterns being significantly different. Importantly the responsibility for applying the camouflage had been shifted from the Base Ordnance Depots to the individual unit. A full set of diagrams showing the disruptive patterns to be used was included with this order. The Sherman tanks of British 6th Armoured Division, which had arrived in Tunisia in early November 1942, came directly from the UK and were almost certainly painted in US Army Olive Drab. A semi-official camouflage scheme consisting of the base coat with a disruptive pattern of Light Mud was authorised for these vehicles and photographs show that Black was also used as a disruptive on the Olive Drab base.

With the conclusion of the fighting in North Africa, and in expectation of the invasion of Sicily, the Light Mud and Black camouflage was supposed to have been universally adopted but there is ample photographic evidence that the Sherman tanks of 6th Armoured Division retained their US Army Olive Drab scheme when they were shipped to Italy.

During the first half of 1943 the regiments of 1st Canadian Armoured Brigade were training in northern Britain and being re-equipped with Sherman V tanks. In May 1943 instructions were issued regarding vehicle camouflage for what was termed '1 Canadian Infantry Division and additional troops'. The Canadian units were at that time preparing for the invasion of Sicily (3). The orders stated that all vehicles were to be camouflaged and drawings were provided which showed how the disruptive colour was to be applied.

The colours to be used were SCC 4 Stone and SCC 14 Black. The former has been described as a dull medium earth brown with a pink hint. The drawings clearly showed that the entire vehicle was to be covered in the new colours and the amounts allocated, indicated by weight in the orders, suggest that the dark colour was to cover approximately one-third of the surface area and was to be applied along

## Notes

1. The SCC system from British Standard (BS) 987c was introduced in 1941 and it should be stressed that although I have included the names here they were not decided upon until some time later.

2. Neither the number of this new General Order, nor its exact date are known and our evidence is based on a directive issued by XIII Corps on 23 April 1943 which refers to an order received from GHQ Middle East Forces. It is often thought that the new colours were chosen in anticipation of a possible landing in Sicily or Sardinia and there is undoubtedly some truth in this assumption. Like Desert Pink, Light Mud was a locally-produced shade.

3. SD 5/15A/43 dated 2 May 43, Command Special Instructions. Appendix A covered vehicle camouflage. This was later clarified by SD 5/15A/43 dated 8 May 43, Special Instructions for Force Y.

## Notes

1. It should be remembered that many of these tanks were taken over from British regiments.
2. A comparison of the tanks' census numbers, where visible, and the shipping lists compiled for the landings confirms that these tanks did indeed take part in the Sicilian operation.

the lower edges of the vehicle. It was not to be extended to the underside which was to be left in 'the original light colour'. The remainder of the order went on to describe the size and colours of the RAF roundel aerial recognition device and the application of unit serial numbers and tactical signs.

That this new camouflage scheme, in some form, was actually applied to the tanks is confirmed by surviving diary entries for the Ontario and Calgary Regiments where the light colour is variously described as a very bright tan, light yellow-brown and sand. The diary of the brigade's workshop is somewhat clearer referring to '14 Black' and '4 Brown'. It would seem that the painting was undertaken during the first week of June 1943. There is also some evidence that the British scheme of SCC 2 Brown and SCC 1A Dark Brown was applied to some vehicles and an example of this is shown on page 18 of the Camouflage & Markings section (1).

Somewhat confusingly, contemporary photographs show Canadian tanks operating in Sicily with what is definitely a two-colour disruptive camouflage pattern and also vehicles that are painted in a single colour (2). It may have been that there was insufficient time to repaint a number of vehicles and the single-colour tanks also seem to lack the RAF roundel. An interesting point to note was that on all tanks the names, unit insignia and tactical signs were all covered with paper, fixed to the vehicle with varnish, before embarkation as a security measure. When the paper was removed a dark surround remained and this can be seen in many photographs.

From early 1943 elements of 4th New Zealand Armoured Brigade and 11th South African Armoured Brigade were in training in Egypt and both formations were equipped with Sherman tanks received directly from North America.

Most of these vehicles, if not all, were repainted in the camouflage scheme of Light Mud and Black but it is possible that the original base coat of Olive Drab was sometimes used as the disruptive in place of Black. Although the prescribed patterns, as given in the April 1943 order mentioned above, were usually followed there is photographic evidence that the disruptive colour was applied in patterns which were

*Above: Examples of disruptive camouflage applied to tanks of 4th New Zealand Armoured Brigade. A. Photographed in Egypt in June 1943 these Sherman III tanks are uniformly painted in accordance with the April directive except for the last vehicle in line which clearly displays the October 1942 scheme using what may be the original US Army Olive Drab base coat as the disruptive. B. A Sherman III of B Squadron, 18th Armoured Regiment with a very unusual disruptive pattern which bears no relationship to any of the official schemes. C. An example of the green and brown camouflage applied to some, but not all, the tanks of 18th Armoured Regiment during the spring of 1944. D. Tanks of A Squadron, 18th Armoured Regiment waiting to cross the Adige River in northern Italy in April 1945. The vehicle in the foreground still has traces of a two-colour disruptive scheme.*

*Above: Examples of camouflage applied to South African and Canadian tanks. A. A Sherman III of The Pretoria Regiment, 6th South African Armoured Division probably photographed during the summer of 1944. This tank is camouflaged in Light Mud with a Black disruptive pattern but there is a third colour visible here, possibly US Army Oliver Drab. This tank is also shown on page 24 of the Camouflage & Markings section. B. The tanks that were received by the South Africans from the latter half of 1944 almost certainly retained their initial coat of US Army Olive Drab as these Sherman IB tanks have. C. Photographed in late January 1944 this Sherman V of C Squadron Lord Strathcona's Horse of 5th Canadian Armoured Division is painted in what is very clearly a single colour which is almost certainly US Army Olive Drab. The darker areas around the turret and next to the RAC flash are the remains of the waterproofing compound which was used on all the tanks when they were shipped from Britain to Italy. D. Tanks of B Squadron Lord Strathcona's Horse lined up for inspection by HM King George VI in July 1943. All the tanks were repainted for the occasion, presumably using SCC 15 which had been introduced in the previous April. There is a mention of the Shermans of 1st Canadian Armoured Brigade being cleaned with petrol at this time which would have given the tanks a very dark appearance.*

similar to that laid down in October 1942 or which bore no relationship to either. In April 1944, by which time both brigades were in Italy, SCC 15 Olive Drab was introduced to replace the SCC 2 camouflage employed in the UK and from this time Lend Lease equipment, which of course included the Sherman tank, was not to be repainted (1).

Towards the end of April 1944 many of the tanks of the 18th Armoured Regiment of 4th New Zealand Armoured Brigade were repainted in a camouflage scheme which was described as green with a brown disruptive and an example of this is shown on page 21 of the Camouflage & Markings section where I have assumed that SCC 15 and SCC 2 were the colours used. Why this was thought necessary is not clear and it was not applied to all the regiment's tanks.

In June all the brigade's tanks were supposed to have been repainted in a green shade which was presumably SCC 15 but vehicles with a disruptive pattern could be seen until at least the end of the year. Many South African Sherman tanks were photographed in their Light Mud and Black camouflage well into 1945 and were probably not repainted until the victory parade held at Monza in May of that year.

Units of 5th Canadian Armoured Division began arriving in Italy in late 1943 and were supposed to take over the tanks of British 7th Armoured Division which was due to return to the UK. Many of these vehicles had served in North Africa and were found to be in such poor condition that few were serviceable. In late November 1943 Major-General Guy Simmonds, the Canadian division's commander, was given the choice of accepting the Sherman III tanks of 7th Armoured Division or waiting for an allocation of new Sherman V vehicles which were expected to arrive via Egypt.

Simmonds opted for the latter, reasoning that replacements for the diesel-powered tanks would be hard to come by, and was promised that the first vehicles would arrive in the following month. But by 19 December 1943 just two tanks had been received and it would be months before the division had its full allocation. These vehicles would certainly have been painted in US Army Olive Drab and photographs of the division's tanks do present a very uniform appearance. Towards the end of 1944 a number of tanks were handed over from British 4th Armoured Brigade and these were still in their Light Mud and Black camouflage.

All vehicles were supposed to be repainted in SCC 15 when the Canadians were preparing to leaving Italy for the Netherlands in early 1945 but photographs exist of Sherman V tanks loaded onto railway cars that are clearly painted in a two-colour camouflage scheme.

## Notes

1. Not to be confused with US Army Olive Drab. While the British colour was similar it tended to fade to a greener hue. The Canadian experience of completely repainting Lend Lease vehicles and the specific mention in the April 1944 directive that it was not required suggests that far more Sherman tanks may have been repainted than was previously thought.

Dragon Models Ltd
B1-10/F., 603-609 Castle Peak Rd,
Kong Nam Industrial Building,
Tsuen Wan, N. T., Hong Kong
www.dragon-models.com

Tamiya Inc
Shizuoka City, Japan
www.tamiya.com

Star Decals
Johan Lexell
Stockholmsvägen 12 B,
15240 Södertälje, Sweden.
www.star-decals.net
Most of the old Bison Decals line have been reprinted under the new name.

Accurate Armour
Units 6 & 7,
Kelburn Business Park,
Port Glasgow,
Inverclyde,
Scotland PA14 6TD
www.accurate-armour.com

Asuka Model/Tasca
3-5-30, Toyoda,
Suruga-ku,Shizuoka-city,
Shizuoka, Japan
www.tasca-shop.onck.net

Royal Model
Via E. Montale, 19-95030,
Pedara, Italy
www.royalmodel.com

Italeri S.p.A.
via Pradazzo 6/b,
40012 Calderara di Reno, Bologna, Italy
www.italeri.com

Hauler
Jan Sobotka,
Moravská 38, 620 00 Brno,
Czech Republic
www.hauler.cz

Voyager
Room 501, No.411 4th Village,
SPC Jinshan District, Shanghai 200540,
P.R.China
www.voyagermodel.com

Griffon Model
Suite 501, Bldg 01, 418 Middle Longpan Rd,
Nanjing, P.R. China
www.griffonmodel.com

Aber
ul. Jalowcowa 15, 40-750 Katowice, Poland
www.aber.net.pl

E.T. Model
I can personally recommend this company for their service.
www.etmodeller.com

Friulmodel
H 8142. Urhida, Nefelejcs u. 2., Hungary
www.friulmodel.hu

Modelkasten
Chiyoda-ku Kanda, Nishiki-Cho 1-7, Tokyo, Japan
www.modelkasten.com
Very difficult to navigate but worthwhile.

Archer Dry Transfers
www.archertransfers.com
The company's website was closed when this book was prepared.

Eduard Model Accessories
Mirova 170, 435 21 Obrnice,
Czech Republic
www.eduard.com

Master Club
www.masterclub.ru
It appears that this firm is closely associated with Armour35, a Russsian mail-order firm.

RB Model
Powstancow Wlkp.29B,
64-360 Zbaszyn,
Poland
www.rbmodel.com

OzArmour
P.O Box 51,
Burpengary, QLD, 4505,
Australia
www.ozarmour.com.au

Kaizen
Like many Chinese companies this is hard to pin down to an exact location. The tracks are available from MS Models Japan, a firm for which I can personally vouch.
www.msmodelswebshop.jp

Passion Models
This Japanese company is affiliated with MS Models Japan.
www.passionmodels.jp

Black Dog
Petr Polanka,
Letecká 549,
Libcice nad Vltavou,252 66,
Czech Republic
www.blackdog.cz

*In writing this book I referred in part to the published works of Jeffrey Plowman for the New Zealand units and Barry Beldam and Steve Guthrie which deal with the Canadians. Both the Canadian and New Zealand governments possess extensive and easily accessible photographic archives, and a great deal of information was gleaned from these. Large parts of the wartime diaries of the Canadian regiments covered in this book are available from Canadiana Héritage, a branch of the Canadian Research Knowledge Network, and these proved to be a mine of information. Official South African wartime documentation, particularly regarding vehicle allocations and troop movement lists, is difficult to obtain but proved to be extremely helpful as were the individual unit histories. I was fortunate enough to be able to consult a number of reliable studies on the contribution of South African and Rhodesian units including* The First South African Armoured Battle in Italy During The Second World War: The Battle of Celleno-10 June 1944 *by Evert Kleynhans and* Red Tabs: Life and Death in the 6th South African Armoured Division 1943-1945 *by James Fraser Bourhill. In regard to the section on vehicle camouflage I must make special mention of the encyclopedic knowledge of British researcher Mike Starmer. I should say that this does not imply Mike's endorsement of my conclusions and we have agreed to disagree in the past. I would also like to acknowledge the contributors to Brett Green's Missing Lynx website and The Sherman Minutia website, particularly my old friend Leife Hulbert, and the modellers who graciously allowed me to publish the images of their work. I must make special mention of Jan Zdiarsky from Eduard Model Accessories, Freddie Leung of Dragon Models and the crew at E.T. Models who are always ready to help. As always, I am indebted to Karl Berne, Valeri Polokov and J.Howard Parker for their invaluable assistance with the photographs and period insignia.*

*Said to have been photographed near Mileto, a small town about 40 kilometres north of Reggio Calabria, this Sherman V is one of the tanks assigned to the regimental headquarters of 11th Canadian Armoured Regiment (The Ontario Regiment). The name War Admiral refers to a famous racehorse of the period. Just above the name and over the WD number, T 146559, handles have been welded to the hull side and a similar fixture can be seen on the transmission cover. Note the white rectangle on the lap gunner's hood. This can also be seen on the gun mantlet of other tanks and may identify a headquarters vehicle.*